Human Genetics

Fourth Edition

A. M. Winchester
University of Northern Colorado

Thomas R. Mertens
Ball State University

Charles E. Merrill Publishing Company
A Bell & Howell Company
Columbus   Toronto   London   Sydney

Published by Charles E. Merrill Publishing Company
A Bell & Howell Company
Columbus, Ohio 43216

This book was set in Clarendon
Text Designer: Ann Mirels
Copy Editor: Mary Pound
Production Coordinator: Rebecca Money
Cover Art by M. C. Escher, courtesy of Gemeentemuseums, Gravenhage, Hague, Holland
Cover Design: Tony Faiola

Library of Congress Catalog Card Number: 82—61526
International Standard Book Number: 0—675—20008—3
Printed in the United States of America
   3 4 5 6 7 8—87 86 85

preface

In the four years since the third edition of this book was published there has been a remarkable expansion of public interest in and awareness of human genetics. Television news reports, specials, and documentaries, as well as magazine and newspaper articles, have all contributed to public awareness of advances in human and medical genetics.

The total volume of scientific knowledge is said to double every ten years, while the volume of biological knowledge doubles every five years. Human genetics knowledge is said to double every two years! If professionals in the field have a difficult time keeping abreast of the many new developments in the discipline, how can laypersons hope to have even a cursory understanding of the many developments that are taking place?

We are convinced of the importance of public understanding of human genetics and we hope that books such as ours will provide a degree of scientific literacy in this discipline. Developments in human genetics affect our lives today and may have an even greater effect in the future. Many of these developments raise serious moral and ethical questions with which we as individuals and as members of society must deal. We must be aware of and understand the basic principles and new developments in human genetics if we are to be wise consumers of the new genetic services available to us.

Although this book can serve as a textbook in a beginning course in human genetics, we believe that it may well be of interest to the general public. We have tried to keep the technical jargon and complexities to a minimum, and to define or explain each bold-faced term as it is introduced in the text. We have attempted to write a book that will be useful to persons with a minimal background in biology, biochemistry, and mathematics. We believe that the book can be employed successfully with college freshmen who have had no previous college biology courses.

The fourth edition continues to use photographs and drawings to enhance reader interest and to make the book more useful as an instructional tool. We have added numerous new illustrations, and credits to the sources of these illustrations are given in the legends accompanying them. We are indebted to these individuals for providing the photographs and drawings.

The fourth edition includes a number of innovations designed to enhance its usefulness as a textbook. Appropriate questions and problems have been developed for each chapter, and answers to these are included in an appendix at the end of the book. Each chapter ends with a selected list of references designed to provide the reader with additional reading resources. References range from the technical to the popular. Liberal use of *Scientific American* articles has been made. A number of changes and additions are reflected in new chapters in the book as well. Separate chapters have been developed on immunogenetics and blood genetics. New chapters on somatic cell genetics and behavior genetics have been added, reflecting the rapid development of these areas. A final chapter concerned with human genetics for the citizen has also been added. The chapter emphasizes developments in human genetics education and the expansion of genetic counseling services. We have also included in this edition V. A. McKusick's Human Gene Map. The index provides the five-digit catalog numbers referring to McKusick's *Mendelian Inheritance in Man*, Fifth Ed. (1978).

We are indebted to the following persons who have read the manuscript and made helpful corrections and suggestions for improvement: Dr. David Fox of the University of Tennessee and Dr. Lawrence Eckroat of Behrend College of the Pennsylvania State University. Dr. Lee Ehrman of the State University of New York at Purchase read the chapter on behavior genetics and made valuable suggestions and corrections that were incorporated into the final version of that chapter. As the authors, however, we assume full responsibility for any errors remaining in the book.

Finally, we are indebted to Karen M. Vincent of Muncie, Indiana for typing the manuscript and for assisting in proofreading. Without her dedicated and conscientious efforts, preparing the manuscript would have been a much more difficult task.

A. M. Winchester
T. R. Mertens

contents

The Physical Basis of Heredity

HOW HAS YOUR LIFE BEEN INFLUENCED BY heredity? In more ways than you may realize. Of course, you easily can see how many of your physical features have come down to you from your parents and more distant ancestors. Heredity, however, goes much further than such obvious characteristics. For example, many of the physiological reactions now taking place in your body are governed by the genes you received. In addition, your susceptibility to disease, your aptitudes, many of your behavior patterns, and even your life span are influenced by these genes. Although your environment has played an important part in the degree of expression of your genes, these genes have given you all your potentialities. The very fact that you are reading these lines as an intelligent human being shows that you received a good combination of genes. Had just one of the many thousands of genes in that one cell from which you developed been different, you might now be a patient in a mental hospital. Another single gene variation could have made you physically disabled, unable to move about under your own power. Still another might have resulted in such abnormal embryonic development that you would never have survived the rigors of intrauterine existence and would have been aborted. Some of your genes, of course, might not be exactly those you would have chosen. You might have one that causes your skin to freckle when exposed to sunlight, or one that has given you soft tooth enamel, or one that has caused astigmatism. However, your overall gene combination must be very good.

Everyone can benefit from some knowledge of genes and how they operate. Not only is it interesting to learn how traits are passed down through the generations and how to predict the appearance of traits in future children, but such knowledge can aid in the recognition and prevention of many serious genetic defects. Many people today are leading normal lives even though they inherited genes for disabling defects, because geneticists have learned how genes cause certain defects and what to do to thwart such abnormal action. A knowledge of heredity can also aid a physician in the diagnosis and treatment of many diseases. Because most of the microorganism-caused diseases that formerly took a heavy toll of human life now have been conquered, more of the human afflictions that physicians are called on to treat have a genetic basis. Medical genetics is becoming increasingly recognized as one of the most important subjects being taught in medical schools today. However, to date, U.S. medical schools have been rather erratic in formalizing genetics requirements for their students.

In this first chapter we shall learn something about the nature of these important genes, how they exert their influence, and how they are carried on the chromosomes of the cells.

The Nature of Genes and How They Work

What is a gene like, and how does it exert its influence on development? Until about the middle of this century, answers to these questions were rather nebulous. Genetic crosses had demonstrated that genes are present in cells, but genes had never been seen; while their final effects could be demonstrated, the way they operated remained a mystery. Great improvements in microscopes and chemical techniques in recent times, however, have made it possible for us to see genes and to analyze their chemical structure down to the finest detail. In addition, we now know much about how they operate.

Discovery of Gene Structure

Since genes are unique and since they control the development of the organism, one might expect to find that they are made of some rare elements not found in ordinary compounds. Actually, they are made of some of the most common elements: carbon, hydrogen, oxygen, nitrogen, and phosphorus. It is the way these elements are put together rather than their nature that gives genes their unique ability to regulate the development, structure, and function of organisms. In 1962 the Nobel prize was awarded to James Watson, an American, and F. H. C. Crick and M. H. F. Wilkins, both Englishmen, for their discovery of the structure of the chemical, **deoxyribonucleic acid (DNA),** that makes up the gene. These scientists developed a model showing that the structure of DNA is in the form of a double helix, somewhat like a long flexible ladder twisted about itself. See Figure 1–1 for a photograph of a gene greatly magnified under the electron microscope.

The Chemical Nature of a Gene   The outer supports of the DNA ladder are made up of deoxyribose sugar molecules alternating with phosphate groups as follows: D–P–D–P–D–P. . . . The connections between these two supports, equivalent to the rungs on the ladder, are made of two types of paired bases. One pair is made of **adenine** and **thymine,** abbreviated **A** and **T**, respectively. The other pair is made of **cytosine** and **guanine,** abbreviated, respectively, **C** and **G**. The DNA molecule thus consists of two complementary strands with the bases of one strand paired with those of the other strand.

How can there be so many different kinds of genes when the variety of parts is so limited? The DNA ladder is very long and contains many paired bases. They form four kinds of steps in the ladder—AT, TA, CG, GC—which may be present in any sequence. The process can be compared to writing a long sentence with hundreds of words using a four-letter alphabet. Since an alteration of one letter can change the meaning of the entire sentence, the different kinds of sentences are almost limitless.

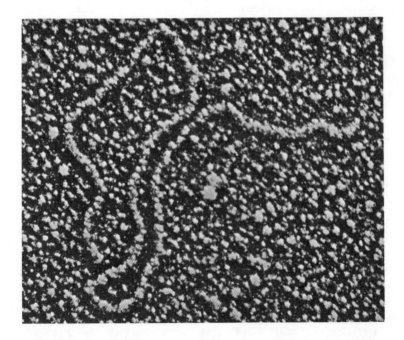

FIGURE 1–1.   A single gene, photographed under the great magnification of the electron microscope, looks like a ladder twisted about itself. (Courtesy Johnathan Beckwith.)

Gene Duplication (Replication)

When the nature of genes was discovered, it became possible to understand one of the great mysteries of life—how genes duplicate themselves. Genes must be able to make exact copies of themselves, a process known as **replication,** or there could be no growth or reproduction. Each time a cell divides, the genes must replicate before being segregated into the **daughter cells,** such that each receives the same complement of genes present in the parent cell.

In replication, a gene first splits down the center, much like a zipper being unzipped. The paired bases are held together by weak hydrogen bonds, and they become disjoined at these bonds. The gene thus becomes split into halves, each with a single strand of sugar-phosphate and the attached single bases. As the splitting is taking place, each half gene attracts to itself new complementary parts that are identical to those it has lost. As a result of replication, two genes are produced, each an exact duplicate of the original (see Figure 1—2). Within a cell there are free bases that become attached to the sugar-phosphate and

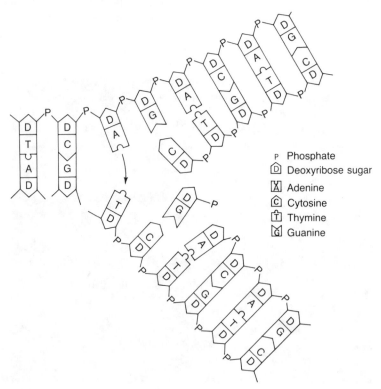

P   Phosphate
Ⓓ   Deoxyribose sugar
Ⓐ   Adenine
Ⓒ   Cytosine
Ⓣ   Thymine
Ⓖ   Guanine

FIGURE 1—2. A small portion of a gene during replication. The DNA separates at the weak hydrogen bonds where the bases join, and each half of the DNA molecule attracts to itself the parts that have been lost.

are thus available to the replicating genes. Adenine on the half-gene attracts to itself thymine, thymine attracts adenine, cytosine attracts guanine, and guanine attracts cytosine. In rare cases, a mistake may be made during replication, and a base of the wrong kind becomes inserted into the replicating gene. The result is a **mutation** (see Chapter 10).

How Genes Work

Genes exert their influence through proteins. Amino acids are assembled into proteins on the ribosomes in the cytoplasm. The genes that determine the structure of the proteins are located in the nucleus, however. How can a gene in the nucleus convey its information to the ribosomes in the cytoplasm? Messages in the form of **messenger-RNA (m-RNA)** are formed by the genes and passed out to the ribosomes. RNA **(ribonucleic acid)** is similar to DNA, but with several characteristic differences: it has a single strand rather than a double strand, the sugar is ribose rather than deoxyribose, and the base **uracil** replaces thymine.

When a certain type of protein is needed in the cell, a stimulus is received by the gene. This stimulus causes the two strands of the gene to separate as if the gene were going to replicate, but instead a strand of m-RNA is produced along one strand of the gene. Complementary bases are assembled with uracil rather than thymine opposite adenine. For example, a sequence on one strand of the gene of TGCATC would lead to the sequence ACGUAG on the m-RNA. The m-RNA then passes from the nucleus to the cytoplasm where its message is translated into protein on ribosomes.

Amino acids are brought to the ribosomes by another form of RNA, **transfer-RNA** (see Figure 1–3). These amino acids are assembled into **polypeptide chains,** with the sequence of amino acids determined by the code in the m-RNA as received from the gene. A protein may consist of one or more polypeptide chains. Some of these are structural proteins, which become part of the protoplasm of the cell and thus contribute to growth and repair. Others are functional proteins, such as enzymes and hormones, which regulate the many reactions that take place within and between cells. The importance of genes can be traced to the many functions of the proteins for which they code.

Proteins derived from food are broken down into their component amino acids in the intestine; these in turn are absorbed by the blood, which then carries them to the cells of the body. Within these cells the amino acids are reassembled into proteins according to the messages from the genes. A slice of beef contains proteins put together according to the genes of a cow; the amino acids released from these proteins are reassembled according to the instructions of human genes to form

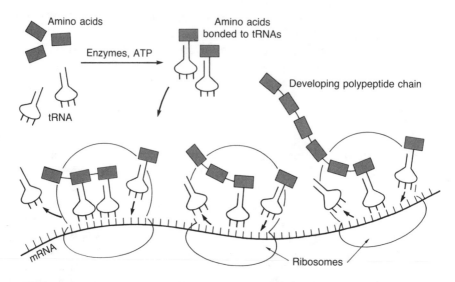

FIGURE 1—3. The method of gene action. Genes (DNA) in the chromosomes in the nucleus transcribe messenger-RNA (mRNA), which passes into the cytoplasm where it directs the assembly of amino acids into polypeptide chains at the site of the ribosomes. Individual amino acids are carried to the ribosome and positioned for assembly in the polypeptide chain by transfer-RNA (tRNA). For further details see also Chapter 10.

human proteins. A pregnant woman and a pregnant cat may eat exactly the same food, yet in one case the elements in the food make a human baby, and in the other they make kittens (see Figure 1—4). The genes in the developing embryos rather than the source of the food elements determine the final product. Of course the food eaten by an expectant mother is still important. The developing baby must have an adequate supply of the different kinds of raw materials needed by its embryonic cells if its genetic potential for development is to be fully realized. A nutritional deficiency can result in a defect even though the baby's genes are normal.

The Variety of Genes

There are many different varieties of each gene. With some exceptions (such as identical twins), no two people receive exactly the same gene combination (see Figure 1—5). Usually, then, no two people produce identical cellular proteins. This fact becomes apparent when we try to transplant tissues from one person to another. When a foreign protein is introduced into the body, antibodies are produced that react with it and reject it. As a result, a tissue or an organ can be successfully transplanted only if treatment is given to suppress antibody production. (See Chapter 7 for details of this type of reaction.)

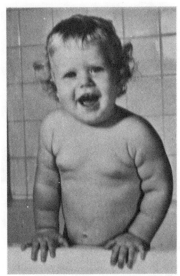

FIGURE 1—4. What a difference genes can make. These two very different bodies were made of the same chemical materials; it was only the difference in the genes in the two fertilized eggs that caused these materials to form a kitten in one case and a human baby in the other.

Children of the same parents share more genes in common than children of different parents. Still, each conception represents a different combination of parental genes no matter how many children a couple may have. Those in the same racial group have more genes in common than those in different racial groups. The differences are more pronounced between members of different species. We share many genes in common with chimpanzees, but the differences are much greater than those between different human races. A transplanted organ, therefore, has the greatest chance of being successfully received when it comes from a close relative.

Some attempts have been made to transplant organs of different species into humans. Success in this would be a great advantage because living human kidneys are not easily obtained, and finding human hearts and livers is even more difficult. Unfortunately, however, the protein difference is so great between species that, even with strong antibody suppression, there is little hope of successful acceptance.

Chromosomes—The Carriers of the Genes

Within the nucleus of a human cell there are many genes, perhaps hundreds of thousands. Yet each time a cell divides, each of these genes must replicate and segregate so that each daughter cell receives one of

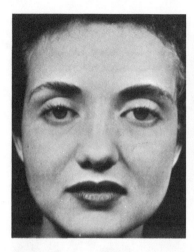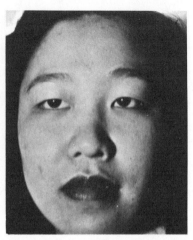

FIGURE 1–5. Variation in facial features are almost infinite. These two young women have features characteristic of their racial background, but each also has unique configurations that make her stand out as an individual like no other in the world.

each kind of gene that was present in the parent cell. Since there are so many genes, this feat might seem impossible, but the process is greatly simplified because the genes are bound together in large groups on bodies known as **chromosomes.** The genes replicate as part of the chromosomes; segregation to daughter cells is thus greatly simplified. The typical number of chromosomes in each cell is the same for all members of a species but varies considerably among different species. A certain fruit fly used widely in genetic experiments has eight chromosomes, an onion plant has sixteen, a mouse has forty, and a human has forty-six. The composition of chromosomes includes certain proteins, as well as the DNA that makes up the genes.

Development of Knowledge of Chromosomes

Early studies of cells did not reveal much detail, because all parts of the cell were of about the same optical density. In the late nineteenth century, however, it was found that certain stains could differentiate parts of cells. The visibility of chromosomes was enhanced when cells were stained with basic dyes. (The word *chromosome* was chosen because some of the basic dyes were of bright colors [*chromos* = "color," *soma* = "body"].) Using such stains on human tissues, Walther Flemming saw human chromosomes in 1882. Figure 1–6 shows drawings he made from his observations. It was not until 1923, however, that a close estimate of the number of chromosomes was made. Using very thin sections of the tubules from human testes, T. S. Painter estimated that there were forty-eight, although he considered the possibility of there being forty-six.

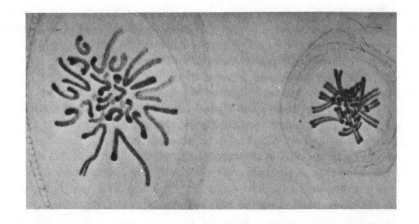

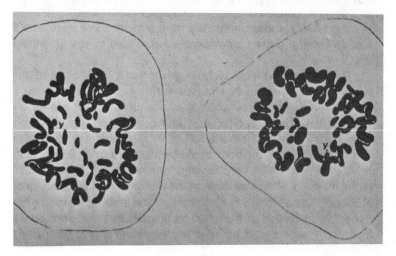

FIGURE 1–6. Human chromosomes as seen by early investigators. The drawings above, made by W. Flemming in 1882, and those below, made by T. Painter in 1923, show that human males have a Y chromosome.

In 1956, J. H. Tjio and A. Levan made an important discovery that opened the way for much more accurate research on human chromosomes. They devised a method of pressing entire cells flat so that all chromosomes in a cell were spread and could be easily counted. They used fibroblasts from lungs of aborted fetuses. (Fibroblasts are rapidly growing cells, and in a given sample, many are in mitosis, which is the time when chromosomes can be seen most clearly [see Chapter 2].) In cells from females the chromosomes could be sorted into twenty-three pairs, with the members of each pair being similar in size and structure. In cells of males, however, only twenty-two matched pairs could be identified, with the two chromosomes of the remaining pair being dissimilar in size and structure. One chromosome of this pair was of medium length, but the other was very short. These are the X chromosome and Y chromosome, respectively, and are related to sex determination. Female cells proved to have two X chromosomes and no Y chromosome (see Chapter 5).

The method of Tjio and Levan was a significant improvement in technique, but it showed only what the chromosomes are like in a fetus that is not viable. Its value would be greatly increased if it could be applied to living people after birth. The development of tissue culture techniques by Theodore Puck in collaboration with Tjio made this possible. A bit of human tissue can be grown in sterile glassware containing a medium supplemented with the nutrients needed for cell growth. Such tissue cultures can be made from almost any part of the body where cell growth and division occurs. Some of the more frequently used tissues are skin, bone marrow from the breastbone, and tubules from the testes. Cells can be removed from a culture and placed on a glass slide where they can be stained and viewed under the microscope. In tissue cultures incubated under optimum growth conditions, many cells will be in mitosis. Studies of the chromosomes from tissue culture cells soon showed that many human defects that had puzzled physicians for centuries were the result of chromosome abnormalities.

The usefulness of the technique was extended when it was found that cultures could be made from blood cells. Circulating blood cells usually do not divide. In fact, mature red blood cells (erythrocytes) have no nuclei and, therefore, cannot divide. White blood cells (leukocytes) do have nuclei, but usually do not undergo mitosis. Some of them, the lymphocytes, can be induced to undergo mitosis, however, if they are treated with an extract of kidney beans. This extract, called phytohemagglutinin, also agglutinates red blood cells so that they can be removed easily. A few drops of blood from a finger prick will give enough cells for several cultures.

The procedure was further improved when it was discovered that the addition of **colchicine,** a drug long used to treat gout, would hold the cells in midmitosis for several hours. When this chemical, which is extracted from the autumn crocus, is added several hours before cells are collected for study, the number of cells with condensed chromosomes is greatly increased. Also, if the cells are placed in a hypotonic solution—one with a concentration of solutes less than that of protoplasm—the cells absorb water and swell. This gives a better spread of the chromosomes when the cells are crushed on a microscope slide.

Identification of Chromosomes

After these discoveries, knowledge accumulated so fast that a standardization of chromosome identification became necessary. Leading geneticists of the world met in Denver in 1961 and devised a system of numbering the chromosomes. The pairs were numbered one through twenty-two according to length, with the XX of the female and the XY of the male designated as the twenty-third pair. This identification is best accomplished when the chromosomes are cut out from a photograph and mounted in order according to length, a process known as **karyotyping.** Figure 1—7 shows the appearance of the chromosomes after they have been made into a karyotype. You will note that each chromosome is double, in the form of two **chromatids.** This is because replication has already occurred in anticipation of segregation in mitosis. Each pair of chromatids is held together at a constriction known as the **centromere.** Since some of the chromosome pairs are of similar length and position of the centromere, it was also decided to group them by letters. (All those in each letter group are of similar length and centromere position.) For instance, the D group includes chromosomes 13—15, which are of medium length and have centromeres near one end; such chromosomes are designated as acrocentric. Group A includes chromosomes 1—3, which are very long and have median, or metacentric, centromeres. Table 1—1 shows the lettered groups and the chromosome numbers and centromere positions included in each.

Improvements in Technique

Although the preceding discoveries had great value in associating many human defects with chromosome abnormalities, a need existed for a means of identifying each chromosome positively. For instance, one defect was known to result from a loss of a part of a chromosome in the D group, but geneticists were not sure whether it was 13, 14, or 15. Hence, it was known simply as the D-syndrome. The stains first used gave the chromosomes a uniform color, but there is variation in the

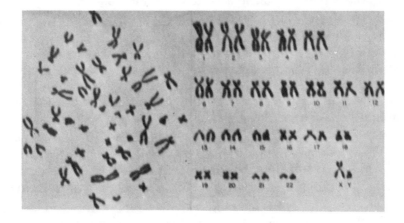

FIGURE 1—7. A karyotype is made by cutting out individual chromosomes from a photograph (left) and arranging them in pairs according to length (right). Each chromosome is double, that is, made of two chromatids, because it has duplicated in preparation for separation during mitosis.

TABLE 1−1.   Identification of human chromosomes

Group Letter	Chromosome Number	Nature of Chromosome
A	1−3	Very long with approximately metacentric centromeres
B	4−5	Long with submetacentric centromeres
C	6−12 and X	Medium length with submetacentric centromeres
D	13−15	Medium length with acrocentric centromeres, with satellites on short arms
E	16−18	Somewhat short with submetacentric centromeres
F	19−20	Short with metacentric centromeres
G	21−22 and Y	Very short with acrocentric centromeres; 21 and 22 may have satellites

distribution of DNA in the chromosomes. A technique was developed that brought out this difference; the chromosomes were stained with **quinacrine mustard,** an acridine dye, and were viewed under a microscope using ultraviolet light. The parts of the chromosomes high in DNA content **(euchromatin)** appeared dark while the parts low in DNA **(heterochromatin)** glowed brightly from the fluorescence of the ultraviolet light. The chromosomes showed alternating light and dark bands, known as **Q-bands.** Since the DNA distribution varies for each chromosome, it became possible to identify each chromosome with certainty by its banding pattern and also to identify individual pieces of chromosomes that may have been displaced in position.

Later, a method was developed that brings out the bands without the need for ultraviolet light. The **Giemsa stain,** long used for blood stains, is added after treatment of the preparation with the enzyme trypsin. The **G-bands** that result are dark where the DNA is low and light where DNA is highly concentrated. Figures 1−8 and 1−9 show how chromosomes differ in appearance when the Q-banding and the G-banding techniques are used.

When **acridine orange** is used as a chromosome stain, the banding pattern is the reverse of that found when the Giemsa stain is used. This is known as the **R-banding technique.** Still other methods bring out **C-bands.** C-banding is especially useful in studying the centromeres.

These new techniques also showed that some chromosomes have small balls known as **satellites** attached to one end. A constriction develops near the end and forms a narrow band that holds the satellite onto the chromosome. Satellites may be seen on the chromosomes of the D- and G-groups.

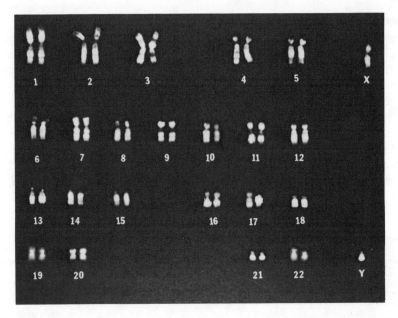

FIGURE 1–8. A karyotype prepared from chromosomes stained with the Q-banding procedure.

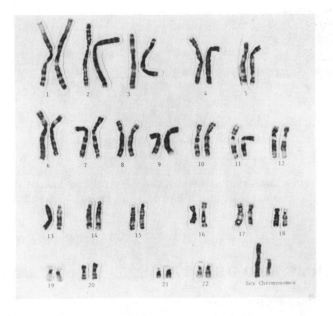

FIGURE 1–9. A karyotype prepared from chromosomes stained with the G-banding technique. The individual from whom the karyotype was prepared is a normal male with twenty-three pairs of chromosomes. (Courtesy Catherine G. Palmer, Indiana University School of Medicine.)

Symbols for Chromosome Bands

Individual bands on the chromosomes may be identified by symbols. The shorter arm of each chromosome is designated as p (from the French "petite") and the longer arm as q. Each arm is then divided into regions numbered from the centromere outward. Bands within each region are numbered, again from the centromere outward. The system is illustrated for chromosome 1 in Figure 1−10. Thus, we use the symbol 1q24 to refer to chromosome 1, the long arm, the second group, and the fourth band in this group. In Chapter 11 we shall learn how these techniques are used to identify even small alterations in chromosomes.

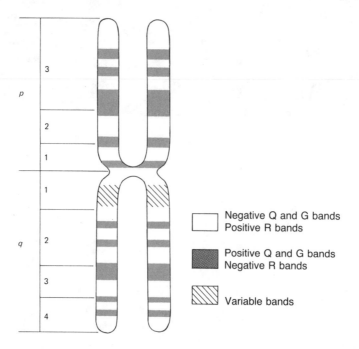

FIGURE 1−10. Method of identification of the regions and bands within the regions of the two arms of chromosome 1 according to a standardized system.

PROBLEMS AND QUESTIONS

1 Suppose that DNA from a certain organism is analyzed and 12 percent of its bases are determined to be adenine (A). Use your knowledge of the Watson-Crick model of DNA structure to determine the amount of thymine (T), cytosine (C), guanine (G), and uracil (U) one would expect to be present in this particular DNA.

2 This chapter describes in a very simplified fashion the relationship between the gene and the characteristic the gene regulates. Place the words *characteristic, protein, DNA,* and *RNA* in proper sequence to symbolize this relationship between gene and characteristic.

3 If P$\overset{D-A}{\diagup}$, P$\overset{D-T}{\diagup}$, P$\overset{D-C}{\diagup}$, and P$\overset{D-G}{\diagup}$ represent the four different "building blocks" (nucleotides) out of which DNA is constructed, where D = deoxyribose, P = phosphate, and A, T, C, and G symbolize the four bases found in DNA, how would the building blocks of RNA be symbolized?

4 Given below is a portion of the base sequence of a gene that controls part of the hemoglobin molecule. Recall that hemoglobin is the red pigment in your red blood cells. Write the messenger-RNA (m-RNA) sequence complementary to this DNA.

DNA code: CAT GTG AAT TGT GGT CTC CTC TTT

5 The m-RNA "message" you determined in Question 4 specifies the amino acid sequence: valine, histidine, leucine, threonine, proline, glutamic acid, glutamic acid, and lysine, where GUA specifies valine, CAC specifies histidine, etc. Now suppose that in a person who has the genetic disease sickle-cell anemia, the DNA code for the same portion of the hemoglobin molecule consists of the following base sequence:

CAT GTG AAT TGT GGT CAC CTC TTT

Locate the "error" in the genetic message for sickle-cell hemoglobin. Indicate exactly how the sickle-cell gene differs from the gene for "normal" hemoglobin (shown in Question 4).

6 Predict the consequence of the change in the DNA code in the sickle-cell gene. That is, how do you think the change in the DNA will affect the amino acid sequence in hemoglobin?

7 A mutation is an alteration in a gene that produces a change in some characteristic of the organism that possesses the mutation. Can you give a more precise definition of a mutation in terms of DNA structure? Review Questions 4, 5, and 6 (and their answers) as a help in answering this question.

8 Place the following words in sequence from simplest to most complex, thereby showing relationships among the terms: organism, DNA, cell, bases, chromosome, nucleus.

9 The chromosomes of humans and those of chimpanzees are very similar in size, shape, G-banding patterns, and in number. Although human cells have forty-six chromosomes and chimpanzee cells have forty-eight, it appears that two chimpanzee chromosomes are almost identical to one human chromosome. What do you think the significance of these observations might be?

REFERENCES

CRICK, F. H. C. 1954. The structure of the hereditary material. *Scientific American* 191(4):54–61.

———. 1957. Nucleic acids. *Scientific American* 197(3):188–200.

———. 1962. The genetic code. *Scientific American* 207(4):66–74.

———. 1966. The genetic code: III. *Scientific American* 215(4):55–62.

HURWITZ, J., and J. J. FURTH. 1962. Messenger RNA. *Scientific American* 206(2):41–49.

KORNBERG, A. 1968. The synthesis of DNA. *Scientific American* 219(4):64–78.

MIRSKY, A. E. 1968. The discovery of DNA. *Scientific American* 218(6):78–88.

SCRIVER, C. R., C. LABERGE, C. L. CLOW, and F. C. FRASER. 1978. Genetics and medicine: an evolving relationship. *Science* 200:946–52.

WATSON, J. D. 1968. *The Double Helix.* Atheneum Publishers. New York, New York.

———. 1976. *Molecular Biology of the Gene,* 3d ed. W. A. Benjamin, Inc. Menlo Park, CA.

YUNIS, J. J., J. R. SAWYER, and K. DUNHAM. 1980. The striking resemblance of high-resolution G-banded chromosomes of man and chimpanzee. *Science* 208:1145–48.

The Bridge of Life

EACH HUMAN LIFE BEGINS AS A TINY SPECK of protoplasm smaller than the period at the end of this sentence. In spite of its small size, however, this cell, known as a **zygote,** contains genes with all the information needed to construct a human body with its many complex organs and functions. Microfilm and computer tapes are gross methods of information storage when compared to the high degree of miniaturization found in the genes.

About one-half of the genes in a zygote come from the **sperm.** A sperm cell is so small that a thousand of them could be lined up across the head of a pin with room to spare. The female contribution to the zygote is an **egg,** which has about ten thousand times the volume of a sperm but still contains approximately the same number of genes. We say approximately because a male actually receives a few more genes from his mother than from his father, while the female receives an equal number from each parent. The reason for this is explained in Chapter 6. Thus, the sperm and egg form the bridge of life that connects the generations of the past with those of the present and that is destined to carry the spark of human life on to the generations of the future. In this chapter we shall learn how genes are passed along from cell to cell and from generation to generation.

Mitosis

Millions and millions of cells make up your body, yet each of these has the same number and kind of genes that were in the zygote from which you came. (Red blood cells are exceptions because they lose their nuclei before they are released into the blood.) This is possible because of the exactness of gene replication and segregation at each cell division. **Mitosis** is the name given to this process. No doubt you have some familiarity with mitosis from a previous biology course, so we need only to point out the major features of the process and show how they are related to heredity. Mitosis is divided into phases for convenience of study.

Interphase

Interphase is actually the period between mitoses. During interphase the chromosomes are very long and very slender, so slender that they usually cannot be seen except with the great magnification of the electron microscope. It is at this time that the genes are exerting their maximum influence. They are producing the m-RNA, and proteins are being constructed. The genes replicate during interphase so that by the time the cell enters prophase both genes and chromosomes are already double but are held together by the centromere, which does not divide until later.

Prophase

By coiling and folding, the long, slender interphase chromosomes become progressively shorter and thicker, so they appear as rodlike bodies in early **prophase.** Figure 2—1 shows a human chromosome photographed with the electron microscope. This highly magnified view brings out the slender threads as they are compacted to make a prophase chromosome. You can see that the chromosome is made of two chromatids. As prophase continues, the nuclear membrane disappears.

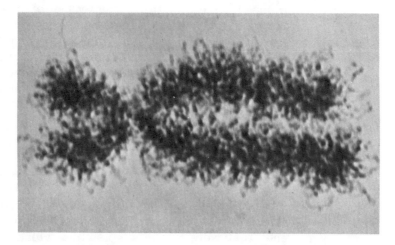

FIGURE 2—1. A chromosome in late prophase, magnified about 100,000 diameters, shows how the slender threads of the interphase chromosomes become coiled and folded so as to greatly shorten and thicken the chromosome. (Courtesy E. J. DuPraw, 1970.)

Metaphase

During the latter part of prophase a **spindle** forms across the cell. It is made of many **spindle fibers,** each of which is a microtubule composed of protein. The centromeres are attracted to the equator of the spindle; once they reach this point the cell is said to be in **metaphase.** Spindle fibers become attached to the centromeres. The centromeres then duplicate, so each chromatid has a centromere. The spindle fibers seem to exert a pull on the centromeres so they separate and move the chromatids to opposite poles of the spindle. As soon as the chromatids are separated, each becomes known as a chromosome.

Anaphase

Anaphase begins as soon as the chromatids are pulled apart and lasts until the newly formed chromosomes reach the poles. Some misunderstanding exists as to the number of chromosomes in a cell during mitosis. Since chromatids become known as chromosomes only after they have separated, the number of chromosomes doubles in anaphase. Hence, the chromosome number is forty-six during prophase and metaphase and becomes ninety-two at the beginning of anaphase.

Telophase

When the segregated chromosomes accumulate at the poles, **telophase** begins. Telophase lasts until the nuclei are restored and the cell has divided. In a sense, telophase is the reverse of prophase. The chromosomes become longer and thinner by uncoiling and unfolding. The nuclear membrane reappears and the spindle breaks down. At the same time a constriction develops at the region of the equator of the spindle and pinches the cell in two. Thus, two genetically identical daughter cells are formed, each with forty-six chromosomes. Figure 2–2 summarizes the main events in the stages of mitosis.

Cycle of Cell Growth and Duplication

When human cells are grown in tissue culture with plenty of food and space, they establish a pattern of growth and duplication that can be divided into stages.

The G_1 Stage

After mitosis the daughter cells enter a period of growth known as the G_1 **stage,** the first growth stage, or gap one. During this stage the chromosomes are fully extended and sending out messages for the synthesis of new protoplasm and enzymes to carry on the activities of the cell. After about eight hours the genes begin replication and

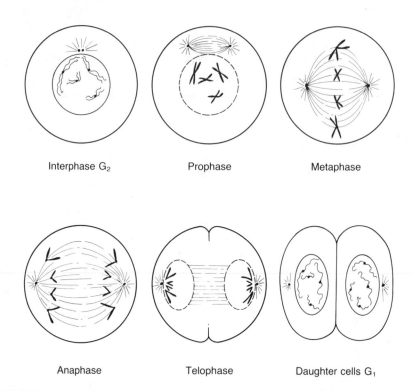

Interphase G$_2$ Prophase   Metaphase

Anaphase   Telophase   Daughter cells G$_1$

FIGURE 2–2. Method of cell reproduction by mitosis. This highly diagrammatic drawing shows only four of the forty-six chromosomes.

growth slows because replicating genes cannot send out their messages (see Figure 2–3).

The S Stage

When the genes begin replication the cell enters the **S stage,** or synthesis stage. This lasts about six hours. Not all growth stops because not all the genes replicate at the same time. The shorter chromosomes finish replicating before the longer chromosomes do. The protein portion of the chromosomes is also duplicated so that at the end of the S stage each chromosome is double, made of two chromatids held together by the single centromere.

The G$_2$ Stage

Once all the chromosomes are replicated, the cell enters **G$_2$,** or the second growth stage. This stage lasts about four hours and the genes are again fully functional. The cell then enters mitosis.

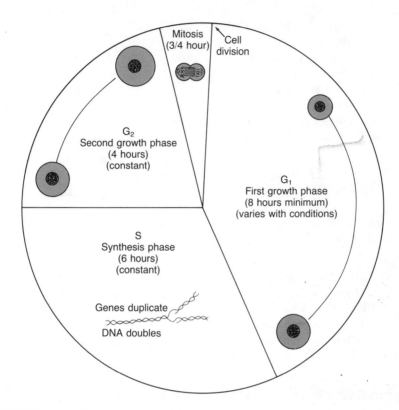

FIGURE 2–3. Cycle of cell growth and duplication under optimum growing conditions.

Variation in Time of Interphase Stages

It is obvious that not all cells in a human body duplicate at this rapid rate. The rate is most rapid in the early embryo, which increases in size about fiftyfold during the first six weeks of life. This rate slows progressively during the rest of fetal life and through childhood. At adulthood it continues only fast enough to replace destroyed tissues and in old age it does not keep up with cell destruction. When not growing at the maximum rate, the time in G_1 is extended while the other stages remain relatively constant. Once the S stage has begun, however, a given cell will continue through mitosis in about eleven hours if it stays alive and healthy. As a person matures, some cells never enter the S stage again. These are said to be in the G_0, or zero growth, **stage.** Because brain cells are in this category, injured brain cells are never replaced by new growth.

What determines when a cell shall enter the S stage? The answer to this question could solve many unexplained phenomena. Cells of your skin may divide at an average rate of once a week, just enough to replace the cells that die and are sloughed off. If you injure your skin, however, this rate is stepped up until the injury is repaired, then the cells return to the slower rate. One theory holds that the injured cells produce a wound hormone that stimulates the cells to enter the S stage. Then, when the cells come together and seal the wound the stimulus stops. Sometimes, however, cell growth and division take place when more cells are not needed. These cells become cancer cells. (See Chapter 14 for a discussion of this important topic.)

An understanding of the stimulating forces might also have valuable applications in the restoration of lost tissue. Cells of the kidneys, heart, and brain are in the G_0 stage. If we could find a way to stimulate their genes to enter the S stage, then growth and restoration of damaged parts might be achieved. We might even use such knowledge to retard the deterioration of tissues in old age.

Some recent studies reveal that regeneration of finger tips and spleens by children under the age of twelve years can occur. This suggests that cells once thought to be differentiated and in the G_0 stage may, under some circumstances, begin active division. Some researchers hold out hope that scientists will learn how to stimulate the human body to initiate mitosis and replace damaged parts much as a salamander can regenerate an amputated limb.

Formation of Reproductive Cells

Mitosis accounts for the continued production of new cells needed for growth and maintenance of the body, but no person can live forever. There must be some way for the genes to be transmitted to new generations if human life is to continue. Sperm and eggs are the reproductive cells that bridge the gap between the generations. A problem must be solved before they are produced, however. If they were formed by mitosis, each would have forty-six chromosomes as do all other human cells formed by mitosis. Hence, a baby formed from the union of a sperm and an egg would have ninety-two chromosomes. Obviously, such a doubling of chromosomes each generation cannot occur. There must be some way to reduce the chromosome number to one-half when sperm and eggs are formed. **Meiosis** is the process that accomplishes this feat. In meiosis there is one duplication of the genes and chromosomes, but there are two divisions of the cells. As a result, each **gamete** (sperm or egg) has only twenty-three chromosomes. Let us consider meiosis in the male first.

Spermatogenesis
If you could look inside a human testis, you would see many feet of threadlike, coiled tubules. A microscopic view of a cross section of one of these tubules will show an outer **germinal epithelium**, which con-

tains cells that will produce sperm. One of these cells divides by mitosis, and one daughter cell remains in place while the other becomes a **primary spermatocyte,** which will produce four sperm as a result of the two meiotic divisions.

A primary spermatocyte goes through the S stage as in mitosis, but we notice an important difference when it enters prophase. At this time, the chromosomes are not only duplicated, they are also paired. Thus, there are twenty-three tetrads made of four chromatids each. In the male there is one exception. The X chromosome is much longer than its mate, the Y chromosome, and the two are paired end to end rather than side by side. (This is shown in Figure 6–1.)

At metaphase of the first meiotic division, the chromosomes line up in pairs in the center of the spindle. The centromeres do not duplicate; two centromeres are already present for each pair. These centromeres pull the paired chromosomes apart and the cell soon divides to give two **secondary spermatocytes.** Each of these has only twenty-three chromosomes, but each chromosome still has two chromatids. The second meiotic division reduces them to single chromosomes. There is no S stage preceding the second division, so the chromosomes enter prophase as they were in telophase of the first division. At metaphase the centromeres duplicate and pull apart the two chromatids of each chromosome. Thus, twenty-three single chromosomes move to each pole during anaphase of the second meiotic division. As the two cells divide, four **spermatids** are produced, each with twenty-three single chromosomes. The two divisions of meiosis are called **meiosis I** and **meiosis II.**

Each spermatid is transformed into a sperm (see Figure 2–4) by throwing off excess cytoplasm (all that is not absolutely necessary to

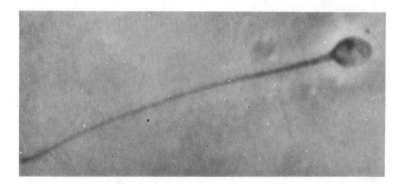

FIGURE 2–4. A living human sperm greatly magnified. Note the light area at the front; this is the acrosome, which contains enzymes needed for the sperm to penetrate the egg. The genes are in the darker area of the head of the sperm.

enable it to reach and penetrate the egg) and by altering its shape. The chromosomes become tightly packed into the head. At the front of the head, vesicles containing enzymes necessary to penetrate the egg accumulate. A tail, needed for swimming, grows out from the other end of the sperm and excess cytoplasm passes down the tail and off at the tip. The **mitochondria,** rodlike bodies in the cytoplasm, accumulate in the middle-piece or neck region of the sperm. The mitochondria produce chemical energy that will be needed to move the tail.

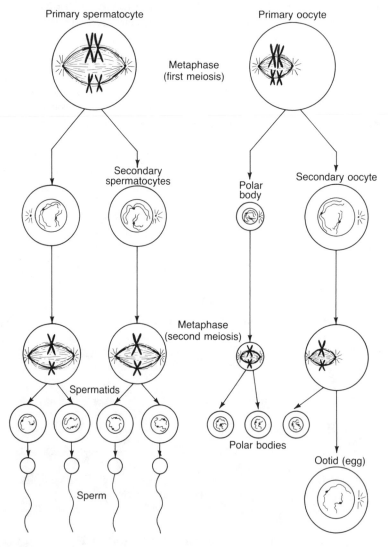

FIGURE 2–5. Meiosis in spermatogenesis and oogenesis. Cell division is unequal in oogenesis, so only one egg is produced by the two divisions. Only four of the forty-six chromosomes are shown.

Oogenesis

Egg production is similar to sperm production, even though the final product is much larger. An egg is larger than a sperm and contains all the food a developing embryo needs until it can begin absorbing food from its mother. Still, the mammalian egg is much smaller than that of birds or reptiles because these latter two must contain enough food to last until the young can begin taking food by mouth.

Eggs are produced in the ovaries. Each ovary has a **germinal epithelium** on the outside that contains cells capable of producing eggs. A **primary oocyte** is formed in this region and then undergoes the two divisions of meiosis. These divisions are unequal. The spindle forms near one edge of the cell, and when the cell divides it forms one large cell, the **secondary oocyte,** and a smaller cell, the **first polar body.** At the second division, meiosis II, a large **ootid,** or **egg,** is produced along with a small **second polar body.** Such unequal divisions allow the egg to receive most of the cytoplasm, including the yolk, that was in the primary oocyte (see Figure 2–5).

A woman usually releases only one egg in each monthly cycle, but a man releases several hundred million sperm at one ejaculation. Such enormous numbers of sperm are necessary because the chance of any one sperm reaching an egg is minute. An egg, however, has a much better chance of being used in fertilization, so the production of relatively few eggs is not detrimental to the reproductive process.

Oogenesis differs from spermatogenesis in another respect. In spermatogenesis the two divisions of meiosis follow each other in close succession. In oogenesis, however, the first meiosis begins before birth but stops in late prophase I and is not completed until the time of ovulation, which will be from about twelve to fifty years later. This long contact of the paired chromosomes may be a cause of chromosome abnormalities, especially in eggs released by women in the latter part of their reproductive life. (See Chapter 11 for details of this topic.)

PROBLEMS AND QUESTIONS _____

1   If a scientist were to determine that a white blood cell of a certain species of animal contained seventy-eight chromosomes, how many chromosomes would you expect to find in each of the following cells in animals of that species?
    (a)   primary spermatocyte
    (b)   secondary oocyte
    (c)   spermatid
    (d)   polar body
    (e)   primary oocyte
    (f)   ovum (egg)

2   Starting with the first growth phase (G_1), place the following stages in correct order to complete the cell cycle: G_1, prophase, S, anaphase, G_2, metaphase, telophase.

3   If 1,983 primary spermatocytes were to complete the entire meiotic process successfully, how many of each of the following kinds of cells would be formed in the course of meiosis?
(a) spermatids
(b) secondary spermatocytes
(c) functional sperm

4   If 1,983 primary oocytes were to complete the entire meiotic process successfully, how many of each of the following kinds of cells would be formed in the course of meiosis?
(a) ova (eggs)
(b) secondary oocytes
(c) polar bodies

5   How many chromosomes are there in a human primary oocyte? How many chromatids? How many tetrads?

6   Which of the following statements are true and which are false?
(a) The four chromatids of a tetrad in a primary spermatocyte are distributed to the four spermatids produced in the two meiotic divisions.
(b) Meiosis could be defined as consisting of two cycles of cell division, each preceded by a duplication of the genes.
(c) Both centromeres and spindle fibers play a role in chromosome movement in mitosis and meiosis.
(d) The two chromatids of a mitotic chromosome should be genetically identical.

7   In what phases of meiosis would you expect each of the following events to occur?
(a) Chromosomes lined up in pairs at the equator of the spindle.
(b) Centromeres duplicate allowing two chromatids of a chromosome to be separated.
(c) Chromosomes undergo pairing to produce twenty-three tetrads.

8   Name the human male cells cited in each of the following descriptions:
(a) The cell in which chromosomes line up in pairs at the equator of the spindle.
(b) Cells formed by the division of a secondary spermatocyte in meiosis II.
(c) The cell that contains only twenty-three chromosomes, but each chromosome consists of two chromatids.
(d) The male gamete.

9 Name the human female cells cited in each of the following descriptions:
- **(a)** The larger cell formed in the first meiotic division.
- **(b)** The cell in which prophase, metaphase, and anaphase of meiosis I occur.
- **(c)** The relatively small, nonfunctional cells formed in both meiotic divisions.

REFERENCES

ANGYAL, J. 1980. Mitotic cell division. *Carolina Tips* 43(5):21–24.

BINKLEY, S. W. 1977. Simulating chromosome behavior. *Carolina Tips* 40(9):33–36.

JOHN, B., and K. R. LEWIS. 1980. *Somatic Cell Division*, 2nd ed. Carolina Biology Reader. Carolina Biological Supply Co., Burlington, N.C.

_____. 1983. *The Meiotic Mechanism*, 2nd ed. Carolina Biology Reader. Carolina Biological Supply Co., Burlington, N.C.

MAZIA, D. 1974. The cell cycle. *Scientific American* 230(1):54–64.

McFALLS, F. D., and D. G. KEITH. 1969. Animal mitosis. *Carolina Tips* 32(1):1–4.

_____. 1969. Animal meiosis. *Carolina Tips* 32(2):5–8.

From Potential to Realization

A SPERM OR AN EGG ALONE IS IMPOTENT, A cell that can live for only a short time and can never achieve the potentialities of the genes it contains. Allow the two to unite, however, and the resulting zygote can bring all these potentialities into realization. A human being, with all its complicated organization and reactions, can be formed. The many physiological and psychological reactions associated with sex are biologically designed to bring a man and a woman together in an intimate association that can make this union of a sperm and an egg possible. In this chapter we shall learn how fertilization is accomplished and how the potentialities of the genes bring about the development of a human being.

Fertilization

When a girl reaches puberty she begins a hormone cycle that results in the release of eggs from her ovaries at a rate of about one every twenty-eight days. Each oocyte develops in a **Graafian follicle** that is small at first but gradually enlarges until it bulges out from the surface of the ovary. Under proper hormonal stimulation, it ruptures and releases the secondary oocyte, which has just been formed by the completion of meiosis I. This is known as **ovulation** (see Figure 3–1). The oocyte is surrounded, along with the first polar body, by a corona of many small follicle cells. It drops into the funnellike opening of the **Fallopian tube,** which practically surrounds the ovary. This tube is lined with cilia that beat in such a way as to carry a thin layer of mucus in the direction of the uterus. The oocyte is caught in this mucus and reaches the

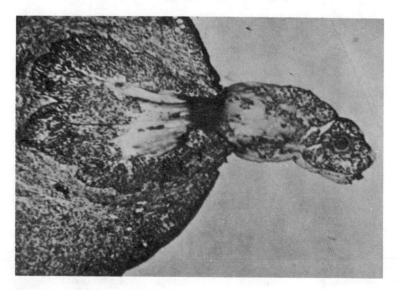

FIGURE 3—1. Ovulation. Like a pimple bursting to release its pus, a Graafian follicle on the ovary bursts and liberates its contents, including the egg that can be seen near the end of the erupting material. (Courtesy R. J. Blandau.)

uterus about five to seven days after ovulation. The secondary oocyte does not undergo meiosis II and form an egg until fertilization occurs. However, since the time it was discovered that mammals produce eggs, it has been customary to call the cell that is ovulated an egg. We shall frequently use the word in this sense even though we now know that it is technically a secondary oocyte.

Semen is deposited in the vagina and the sperm it contains must journey into the uterus and up the Fallopian tube to meet the egg if fertilization is to take place. The egg remains receptive to sperm for only about twenty-four hours after it is ovulated, so sperm must reach it while it is still near the ovary. Sperm swim against the flow of mucus as it moves toward the vagina. This leads them toward the newly ovulated egg (see Figure 3—2). Also, semen contains hormonelike substances, **prostaglandins,** that stimulate the walls of the uterus and Fallopian tubes to undergo wavelike peristaltic contractions that move toward the ovaries. These motions may propel some of the semen toward the egg.

As long as sperm remain in the ejaculatory ducts of a man, they are inactive and survive for several weeks, but during ejaculation they are mixed with secretions of the prostate gland and seminal vesicles that activate them, forming semen. Upon ejaculation they swim vigorously until they exhaust their meager energy supply. At room temperature

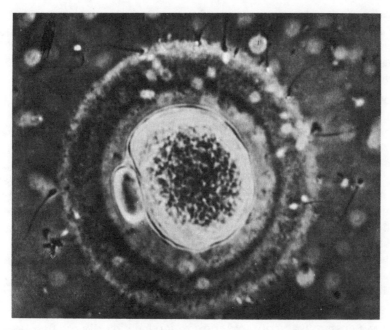

FIGURE 3–2. A living human egg surrounded by its corona being penetrated by sperm in the process of fertilization. Note the polar body at the left. (Courtesy Landrum Shettles.)

this will be within a few hours, at which time they lose their power to achieve fertilization. Placed in a refrigerator, however, sperm move slowly and therefore can accomplish fertilization as long as a week later. If semen is quick-frozen and held at a temperature of $-196.5°C$, sperm retain their ability to achieve fertilization for many years. Cattle ranchers have long used this technique to preserve semen from fine bulls. Now the technique is being used to preserve human semen. Sperm banks are being established throughout the country. These banks allow a man to leave some of his semen as fertility insurance. He can then have a vasectomy and still produce children in the future if he so desires. Sperm banks are also used by some men in occupations where exposure to certain agents might cause harmful genetic changes in their sperm. In cases where the husband is sterile and a couple wishes to have a child that at least is partly theirs biologically, the wife may be artificially inseminated using donor semen (AID). This procedure is being increasingly used in our society as an alternative to childlessness or adoption. In the 1960s the noted geneticist H.J. Muller actively promoted preserving the semen of great men so it would be available to women who might want children with genes from these men. In early 1980 the popular press widely publicized the establishment of a "Nobel laureate sperm bank" in California. One Nobel laureate openly admitted having placed semen samples in this sperm bank. The first child produced as a result of artificial insemination using semen from this sperm bank was a female born in April, 1982. The semen donor was described only as an "eminent mathematician" and university professor.

Sperm can live for several days within the female genital tract because they can absorb some nourishment from the mucus. Sperm may be in the upper part of the Fallopian tube and ready to fertilize the egg upon ovulation. However, ovulation must be within about three days of insemination for fertilization to occur. After this, the sperm may be capable of feeble wiggles, but they have lost their power to penetrate the corona around the egg. The cells of this corona are held together by an intercellular cement and the enzymes in the acrosomes of the sperm are needed to dissolve this cement sufficiently to allow penetration. The enzymes of many sperm are needed for this and, while only one sperm is required for fertilization, many must be present to make way for this sperm. For this reason, a man will be of low fertility, or even sterile, if he has a low concentration of sperm in his semen or a reduced quantity of semen.

Once a sperm head contacts the egg surface, it adheres firmly. Each species of animal produces eggs that will attract and hold sperm only of its own or closely related species. Thus, hybrids can be produced from closely related species, but not from those more distantly related. Domestic cattle can be crossed with the American buffalo to yield a hybrid cattalo, which is sterile because of differences in the chromosomes of the two parent species. The mule is a hybrid of the donkey and the horse, but it is also sterile. No hybrids involving humans and another species have ever been verified.

Once a sperm is trapped, the outer surface of the egg bulges out and engulfs the sperm much like an amoeba engulfs a food particle. The entry of the sperm into the egg stimulates a chain reaction that leads to the release of enzymes that start the development of an embryo. The egg now ceases to trap sperm, and any that may have been already trapped, but not engulfed, will be released. This avoids multiple fertilizations. In rare cases two sperm may be engulfed at about the same time but one usually disintegrates. If both persist, an abnormal type of mitosis occurs that results in such aberrant chromosome distribution that the cells do not survive.

Tubular Development

The sperm entry that causes the enzymes to be released results in an increase in metabolism as indicated by an increased oxygen consumption. The outer membrane around the zygote becomes thickened but is more permeable to oxygen and molecules of nutrients. The sperm head enlarges by absorption of moisture and migrates toward the center of the cell where it unites with the egg nucleus. The cell enters the S stage, and in about thirty hours after fertilization the first mitosis and cell division is completed. The embryo is now made of two cells that are held together within an outer membrane, known as the **zona pellucida.**

An unfertilized egg has a haploid set of genes along with all the food needed to start a new life and normally will not function unless stimulated by the entry of a sperm. However, in some species of animals, such as honeybees, unfertilized eggs will develop into embryos and eventually into haploid adults (drones). Such development of an egg without a sperm is known as **parthenogenesis.** Parthenogenetic diploids are also common in some insect species. Claims of parthenogenic origin of a human fetus have often been made, but these are usually easily disproved. For one thing, any parthenogenetic human would have to be a female. If a child has any characteristics not traceable to genes possessed by her mother, then it is obvious that a father must have been involved. In cases where this test is not conclusive, skin grafts can be attempted. A parthenogenic child should be able to give a successful skin graft to her mother because she could have no genes her mother did not possess. Differences in genes result in differences in the proteins of the skin, causing rejection and therefore proving the existence of the father.

The cells continue to divide and after about three days the embryo consists of about sixteen cells clustered into a small ball, the **morula.** When the embryo reaches the uterus, about seven days after fertilization, it is a hollow ball of cells known as a **blastocyst.** This ball contains an **inner cell mass,** which will form the embryo proper. The rest of the ball will form the embryonic membranes.

Implantation

The embryo faces a critical period when it passes into the uterus. It must become implanted in the wall of the uterus or it will continue its downward movement and pass from the woman's body. At least half of the young embryos fail to become implanted. In a way, this is good because many of these are defective blastocysts and would not develop into normal babies. In many cases, however, implantation may fail because the uterus is not in a receptive state. Some women are sterile for this reason.

Normally, a woman's monthly hormone cycle brings a rich supply of blood to the wall of the uterus, which swells and becomes somewhat sticky at the time when a blastocyst would arrive if fertilization had occurred. The descending blastocyst may then adhere to this wall and begin producing enzymes that digest the surrounding area, creating a small cavity in which it is contained. This is called **implantation.**

At about the tenth to eleventh day after fertilization the outer layer of the blastocyst sends fingerlike projections out into the wall of the uterus and begins absorbing nourishment from the blood of the mother. At the same time it begins releasing a hormone that alters the monthly cycle of the woman so that the lining of the uterus will not be shed as it normally would be in the process of menstruation at about the fourteenth day after ovulation. This is another critical time for the embryo. Many women are sterile, even though they have frequent implantations, because they fail to respond sufficiently to this hormone to prevent the onset of menstruation. Once this time is passed successfully, the chance for survival of the embryo is greatly improved.

The genes in the cells are now very active in sending out their messages for the production of new protoplasm. This brings about rapid growth and differentiation of the cells. Some form membranes around the embryo. One membrane, the **amnion,** surrounds the embryo and is filled with a fluid in which the embryo floats. An embryo is very soft and could be injured easily were it not for this protective fluid. Another membrane forms the **placenta** in close connection with the uterine wall; the placenta absorbs food and oxygen from the mother's blood while giving off carbon dioxide and other waste products into the mother's blood. An **umbilical cord** connects the embryo with the placenta.

Timetable of Development

The organs of an embryo develop according to a definite timetable. Each organ must develop at the proper time and place if it is to be normal. The time and sequence of this development is given below by age after fertilization.

Two Weeks   Embryo is about one-sixteenth inch (1.5 millimeters) in length and is firmly implanted in the uterus. The mother misses her menstrual period and suspects pregnancy.

Four Weeks   Embryo is one-fifth inch (5 millimeters) in length. Differentiation of cells into tissues of the eyes, brain, spinal cord, lungs, and digestive system has begun. The heart is beating firmly and pumping blood through blood vessels to all parts of the body (see Figure 3–3).

Six Weeks   Embryo is one-half inch (12 millimeters) long and limb buds have pushed out with a beginning of differentiation into fingers

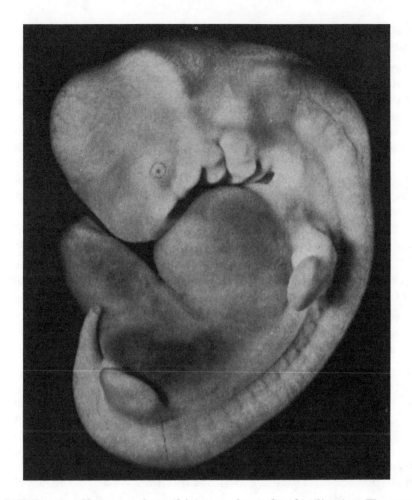

FIGURE 3–3. Human embryo thirty-two days after fertilization. The genes have already constructed the basic parts of a human body. The large swelling under the head contains the heart, which must develop early so it can pump blood through the umbilical cord to the placenta. (Courtesy James Ebert.)

and toes. The head makes up about a third of the body length. Many organs now enter their most vital period of development and are very susceptible to damage. Serious deformity may result if anything interferes with development at this critical time.

Eight Weeks About an inch (25 millimeters) long, the embryo begins to be recognizable as a human being. The eyes are well developed but the lids are stuck together as a protection against probing fingers when muscular movements start. Nostrils are present but are plugged with mucus. Sexual organs have developed to the point where sexual differences can be recognized. Calcium deposits develop in the skeleton to begin bone formation. It is customary to begin calling the embryo a fetus at this time.

Twelve Weeks   Three inches (75 millimeters) long and weighing about an ounce (28 grams), the fetus begins inhaling through its nose but can take in only the salty amniotic fluid. This action, however, develops the muscles that will be needed when air must be inhaled to sustain life after birth. The muscles are now sufficiently developed to permit movement of the arms and legs, although it may be another week or two before the mother is aware of such fetal movements. The body organs are so well formed that the fetus is not as susceptible to injury from drugs, diseases, and other agents as it was in earlier stages. From this point until birth, growth is primarily in size, not in differentiation.

Sixteen Weeks   The fetus is over 5 inches (125 millimeters) long and begins to be cramped in its restricted quarters, so must double over into what we call the "fetal position." The skin is very thin and superficial blood vessels stand out clearly on the surface.

Twenty Weeks   The fetus has now reached about 7 inches (175 millimeters) as measured from the upper tip of the head to the lower part of the rump region.

Twenty-eight Weeks   Now at about 10 inches (25 centimeters) and 2.5 pounds (1.13 kilograms), the fetus reaches an important milestone. It now has a chance to survive if born prematurely. Body metabolism is still not sufficient to maintain a normal temperature, and the chest is still not muscular enough to provide normal respiration, but if the baby is put in an incubator and on a respirator, it has a chance of surviving. The skin is covered with a whitish coating, the **vernix**, which protects it from the continuing immersion in the amniotic fluid.

Thirty-eight Weeks, 266 Days after Conception   The time for normal birth has arrived. The head, which has been pointing up, begins to turn down and presses against the cervix of the uterus. This pressure helps to stimulate the hormone changes in the mother's body that lead to labor contractions, which will expel the baby. It is customary to measure pregnancy duration from the time of the last menstrual period, and this would be forty weeks or about nine calendar months. The fetus, however, is two weeks younger than this.

Teratogenic Agents

Anything that affects a developing embryo in such a way as to cause it to be defective is known as a **teratogenic agent.** Such agents operate in various ways. Some slow growth by inhibition of nucleic acid synthesis and thus prevent the normal progress of mitosis. When this occurs at a critical time in the development of an organ, that organ may never complete normal formation. Some agents cause mutations of genes or alterations of chromosome arrangement in a body part, which then does not develop normally. Other agents may interfere with enzyme function in the cells. An infection of the mother may allow viruses or bacteria to pass across the placenta, infect the embryo, and play havoc with developing organs. A great excess or deficiency of some substances that are a part of normal nutrition can alter growth patterns and result in defects. Also, antibodies from the mother may be teratogenic when they enter the embryo and react with its antigens (see Chapter 7).

The virus of **rubella,** or German measles, was one of the first teratogenic agents recognized. When a woman in early pregnancy has the disease, it causes her only minor inconvenience, but her child is very likely to be born with defective hearing, sight, or heart function. The particular defect that appears is dependent upon which organs are in a critical stage of development when infection occurs. We now vaccinate children for rubella, not so much for their protection, but to prevent them from carrying the infection home to their mothers who might be pregnant. We also make sure adolescent girls are immunized so they will not have an infection if they become pregnant.

Another teratogenic agent came to light in 1961 when many babies in Europe were born with only stumps for arms or legs, a condition known as **phocomelia.** This abnormality had been observed before, but never in such great numbers. It was found that nearly all the mothers that bore such babies had taken the drug **thalidomide** to relieve the nausea of early pregnancy. Tests on experimental animals and volunteer humans had shown no harmful effects, but possible damage to embryos had not been tested. Because the testing program had been limited, this drug was not approved for use in the United States. This experience shows that agents with only minor effects, or no noticeable effects, on adults may cause serious defects in embryos. Embryonic tissue is growing much more rapidly and, therefore, is much more susceptible to damage from substances that alter the growth pattern.

Many agents are now known to be, or are suspected of being, teratogenic. The list below will give an idea of the varied nature of these agents.

Physical Agents

Ionizing radiation, such as X rays or radiation from radioactive isotopes, can be teratogenic in exposures too low to have any noticeable effect on the mother. Since we have learned this, most physicians have discontinued any extended radiation of pregnant women and many even avoid the comparatively small exposures required for X-ray pictures unless absolutely necessary. As an extra precaution, some restrict extensive exposures of any mature woman to the first ten days following her menstrual period because this is the only time when pregnancy is almost certainly not present. Female technicians who work with X rays and radioactive isotopes are usually shifted to other duties as soon as pregnancy is suspected. The **microwaves** from those convenient microwave ovens are even suspected of being teratogenic if a woman stays near them while they are operating for extended periods, but there is no firm evidence for this as yet. `

A reduced oxygen supply, **hypoxia,** such as might come during a journey to a high mountain, can be teratogenic. High-altitude air travel is no longer a problem because of pressurized cabins. An elevated body temperature, **hyperthermia,** can also be teratogenic. The high heat can come from the fever of certain infections. Some birth defects have even been traced to extended stays in hot saunas.

Drugs and Chemicals

In addition to thalidomide, which is no longer available, this category includes some **cancer therapy agents** that function by inhibiting the growth of cancer tissue. Some of the **antibiotics** (especially tetracycline), **antihistamines,** and **immunosuppressive drugs** are also on the list. Even the common **aspirin** is suspected when taken in large quantities, as it might be when used to relieve the pain of arthritis. The steroid hormone **cortisone** has medical value in the treatment of arthritis and certain allergies but should not be used during early pregnancy because it is teratogenic. Other hormones that may be teratogenic are **insulin, androgens, estrogens,** and **vasopressin.** The synthetic hormone **diethylstilbestrol, DES,** greatly increases the incidence of vaginal cancer in girls born to mothers who received the drug in early pregnancy. Even **vitamins A** and **D,** when taken in great excess, are included. The extensive use of **alcohol** during pregnancy may result in a particular type of defect known as the fetal alcohol syndrome. Children with this syndrome have characteristic facial features and are mentally deficient (see Figure 3–4). Many children of mothers who are chronic alcoholics have this syndrome. The mind-altering drug **LSD** and **caffeine,** a drug in coffee, tea, and some soft drinks, are also suspected of being teratogenic.

Many chemicals used industrially are teratogenic, posing a hazard to pregnant women who work with them. These include salts of certain metals, such as **mercury, lead, lithium,** and **selenium.** Fumes from **hydrocarbons,** such as gasoline, benzine, and similar compounds, can be teratogenic when inhaled for long periods in closed areas. Other lesser-known industrial chemicals are included. The **herbicides** used to

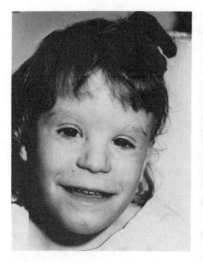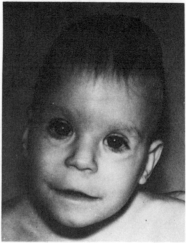

FIGURE 3–4. The fetal alcohol syndrome. The mothers of these two children consumed alcoholic beverages excessively during early pregnancy. Both children are greatly retarded mentally and have facial features characteristic of this syndrome. (Courtesy James W. Hanson.)

defoliate the jungles in Vietnam caused many birth defects in children of women who were near the defoliated regions. Some **pesticides** are in this category as well.

Diseases

Rubella, cytomegalic inclusion disease, toxoplasmosis, syphilis, mumps, and **influenza** are common diseases that are teratogenic. In recent years the public has also been alerted to the danger of **venereal herpes virus** infection to the newborn. Some of these infectious agents infect the embryo and some do their damage through the high fever induced in the mother. Some physicians consider an infection by these diseases during early pregnancy justification for recommendation of therapeutic abortion.

Deficiencies

Too little of vital substances hampers normal development. Deficiencies of amino acids, vitamins, and minerals are examples.

With so many teratogenic agents in our environment, what is a woman to do when she finds that she is pregnant? She might fall back on the argument that her grandmothers and more distant female relatives had mostly normal children and didn't even know there were teratogenic agents, so why should she worry? However, most of these agents did not exist in past generations. Although she may sometimes find it inconvenient, the pregnant woman of today is well advised to

avoid exposure to any agents that are even suspected of being teratogenic, especially during the critical first third (trimester) of embryonic
growth. This precaution will increase the chance of a healthy, normal
child.

Genetic Counseling

As we have just stated, birth defects may be caused by teratogenic
agents. The majority, however, have at least some basis in heredity,
and many of these can be prevented or alleviated if they are discovered
and treated in time. Through genetic counseling, prospective parents
can be informed about the probabilities of defects and what can be done
if they occur. If the risk is very high, some couples may choose not to
have children of their own conception.

Most couples anticipating childbearing can benefit from genetic
counseling. They may not realize that some defect that has appeared in
their ancestry is inherited and may be passed to their children. Many
do not realize the hazards of certain agents during early pregnancy
and should be informed of them. Counseling is particularly important
in some cases; for example, consider the following four cases: (1) A
couple has one child with **cystic fibrosis** and wants to know the risk
involved for future children. (2) A woman planning marriage has a
brother with **hemophilia,** the bleeder's disease, and wants to know if
this affliction might be passed to her children. (3) A forty-two-year-old
woman finds that she is pregnant and has heard that the chance of
bearing an abnormal child increases as a woman grows older. She is
considering abortion, but first wants to know the odds for a defect and
if anything can be done to reduce these odds. (4) A man wishes to
marry his first cousin and wants to know if there is an appreciable
increase in birth defects for children of parents with this degree of
relationship. As we continue our study we shall point out how counseling can have value in specific circumstances.

Genetic counseling has become much more precise with the development of various techniques of prenatal detection of many types of birth
defects. Whereas counselors formerly could only establish a degree of
risk based on family pedigrees, they can now often inform the parents
definitely if a fetus will become a defective child. **Amniocentesis** is the
most widely used technique that makes this possible (see Figure 3–5).
A small sample of the amniotic fluid in which the fetus floats is
removed through a needle inserted through a woman's abdominal and
uterine walls. Chemical tests of this fluid can reveal abnormal metabolic products. Cells in this fluid have sloughed off from the embryo
and can be cultured and analyzed for an inability to produce certain
vital enzymes. Also, the chromosomes of these cells can be karyotyped

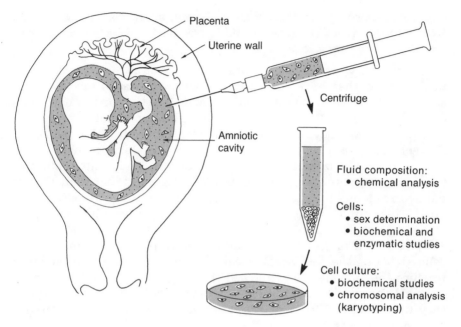

Placenta

Uterine wall

Centrifuge

Amniotic cavity

Fluid composition:
  • chemical analysis

Cells:
  • sex determination
  • biochemical and enzymatic studies

Cell culture:
  • biochemical studies
  • chromosomal analysis (karyotyping)

FIGURE 3—5. Amniocentesis makes possible prenatal detection of genetic and chromosomal defects. (Diagram courtesy Maria R. Arbona.)

and may show abnormalities in structure or number. Stephenson and Weaver (see References) document 182 fetal conditions that have been diagnosed prenatally using amniocentesis or other techniques of pre-natal detection. Of course, prenatal detection has limited value unless the parents are willing to have an abortion if a serious defect is found. Moral issues are involved. Some people feel that we have no right to terminate a pregnancy regardless of circumstances. Others feel that each fetus has a right to be born normal, and that it is best to interrupt a pregnancy that will produce an extremely abnormal child. Some medical malpractice suits have been filed and won by parents who had defective children, claiming that they were not informed about the availability of amniocentesis and therapeutic abortion.

Some groups opposed to abortion have taken strong stands against amniocentesis. Other groups, such as the March of Dimes Birth Defects Foundation, have contended that in 96—97 percent of all amniocenteses, the fetus is found not to have the defect under investi-gation and the prospective parents have their minds put at ease and do not have to face the prospect of terminating the pregnancy. In fact, one can argue that amniocentesis may actually prevent abortions, since prior to the advent of this procedure pregnant women who knew they were at risk for producing a child with a defect such as the Down syndrome terminated the pregnancy without knowledge of the presence or absence of the abnormality. Today, such women, who learn from amniocentesis that their child is free of the expected defect, will allow the pregnancy to go to term.

An example will show how amniocentesis can be of value. A couple has relatives that have had children with **Tay-Sachs disease.** In a screening (detection) test, the couple learns that they are both carriers of the gene for Tay-Sachs disease. Although they do not have the disease themselves, they are at risk for having children with the disease. Such children are normal at birth but cannot produce an enzyme necessary to metabolize certain fatty compounds. The accumulation of these compounds around nerve cells causes a gradual reduction of motor and mental function, with death coming within a few years after birth. This couple had about decided not to have any children, for fear of having one with the disease, until they heard about amniocentesis. They started a pregnancy and had an analysis of the amniotic fluid cells at about the fourteenth week. It showed that the cells could produce the enzyme. They allowed the pregnancy to continue and a normal baby was born. Had the test shown a deficiency of the enzyme, they were resigned to an abortion and a start of another pregnancy with similar tests.

Multiple Births

During the latter part of a woman's pregnancy her physician may use a stethoscope to detect the fetal heartbeat. Instead of one beat, however, the physician may discover two. This is the way a woman often learns that she is among the one percent of expectant mothers in the United States who will give birth to twins. In much rarer instances, three, four, or more heartbeats may be detected. Sonograms are now increasingly used to confirm the presence of multiple embryos. Twins evoke interest because some of them are so much alike that it is difficult to tell one from the other, while others are quite different and may even be of different sexes. This is because twins originate in two very different ways.

The Two Types of Twins

Twins that are very similar in appearance are usually monozygotic, a name given because they originate as one zygote that forms two embryos. Because they look so much alike, they are also known as **identical twins** (see Figure 3–6). Even though monozygotic twins have the same gene combinations, they exhibit some differences because of variation in environment both before and after birth. One may have a more favorable position and a better blood supply in the uterus and will be larger and healthier than the other. One may have a severe infection after birth that slows growth while the other escapes. In very rare

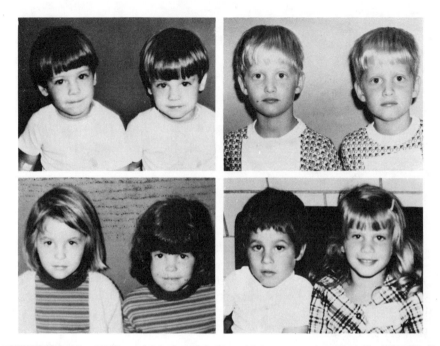

FIGURE 3–6. Different kinds of twins. At top are monozygotic (identical) twin boys and girls. At bottom are dizygotic (fraternal) twins of the same sex and of opposite sexes. (Courtesy Joe Christian, Indiana University School of Medicine.)

cases a gene mutation may occur in one soon after the split. One of a pair of monozygotic twins was found to be an albino while the other had normal pigmentation, apparently as a result of a mutation after separation. Monozygotic twins are of the same sex because the chromosomes that determine sex are the same in both. A very few cases have been found, however, when a chromosome aberration occurred in one after separation and caused a sex difference.

If the embryo separates in the two-cell stage, or anytime through the morula (about the sixteen-cell stage), two individuals are formed and each will have its own set of embryonic membranes. One of these is the **chorion,** the tough membrane around the amnion, part of which forms the placenta. Hence, twins that form in this way are called **dichorionic twins.** After about four days the zygote will have formed a hollow ball, the blastocyst, with an inner cell mass that will produce the embryo. This inner cell mass may divide into two parts and identical twins result, but they will share the same chorion because this membrane is formed from the outer ball of cells that encloses the cell masses. Thus, these twins will be **monochorionic.** About 70 percent of identical twins are monochorionic, a fact easily determined by an examination of the membranes at birth (see Figure 3–7).

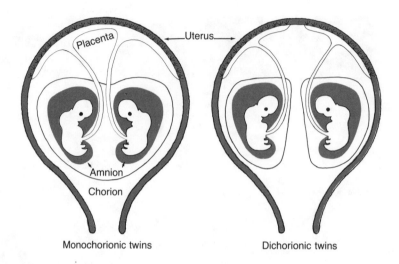

Monochorionic twins   Dichorionic twins

FIGURE 3–7. The two types of twins according to their fetal membranes. Only monozygotic twins can have a common chorion, but both types may be dichorionic.

Dizygotic, or fraternal, twins are produced when two eggs are ovulated at about the same time and both are fertilized. A hormone released along with the egg at ovulation inhibits the action of the ovulation-stimulating hormone from the pituitary gland located at the base of the brain. As a result, further ovulation is usually suppressed. Sometimes, however, a second egg is ovulated before the inhibition is fully effective. When both eggs are fertilized and implanted they produce dizygotic, dichorionic twins. Such twins will be no more alike genetically than brothers or sisters born at different times and they can be of opposite sexes. Sometimes, however, two brothers or two sisters born at different times may have received genes that give them a similar appearance. Since this could also be the case with twins, additional information is needed when it is important to know for certain if twins are monozygotic or dizygotic. If the physician examines the membranes at birth and finds that they are monochorionic, then it is evident that the twins are identical. Dichorionic twins, however, could be of either type and further tests are needed. Since blood exhibits many inherited characteristics, a comparison of the blood may be definitive. If even one of these traits proves to be different, then the twins must be dizygotic. **Dermatoglyphics,** a study of hand- or foot-prints, is sometimes used. The print patterns are not the same for monozygotic twins, but the variation averages no greater than the variation between a person's right and left hands or feet. Dizygotic twins have a greater dissimilarity. Since the discovery of **chromosome polymorphism** (see Chapter 2) we have had another tool. Even one difference in chromosome structure rules out monozygosity, with the possible exception of rare chromosome aberrations in one of the twins. Finally, we can turn to the **skin graft test,** mentioned earlier in this chapter. Tiny bits of skin can be successfully grafted between monozygotic twins, but will be rejected by dizygotic twins.

Most other multiple births are a result of multiple ovulation and are fraternal, but a few are monozygotic. The most famous identicals were the Dionne quintuplets, born in a rural region of Canada in 1934. In this case a single zygote evidently split into five, or perhaps more, parts and five survived. Fraternal multiple births are becoming more common with the increasing medical use of the fertility hormone. Some women are of low fertility because of irregular or suppressed ovulation and this hormone is administered to stimulate ovulation. It is difficult, however, to administer just the right amount to stimulate one ovulation, so two or more frequently occur.

Heredity and Twinning

Is a woman more likely to have twins if some of her close relatives have had twins or if she has already had twins? Since heredity influences the hormone balance and any environmental factors involved are more likely to be similar among close relatives, the answer is yes for fraternal twins. If she already has a pair of fraternal twins, her chance in future pregnancies is about double the average for her population group. If large numbers of relatives have had twins, her chance is greatly increased. There are some families in which most pregnancies result in multiple births. The tendency to multiple ovulation can be transmitted from the males in the ancestry as well as from the females. Identical twinning, however, appears to be due to a developmental accident in most cases, although several families have been reported with a high frequency of monozygotic twins, indicative of a dominant mode of inheritance. The percentage of identical twins born in different parts of the world is about the same, a further indication that heredity is usually not involved. Whatever genes are involved in causing multiple ovulations, however, vary, as the following examples of fraternal twin frequency indicate.

American Caucasians	6.6 per 1000 births
Japanese	2.5 per 1000 births
Nigerians	39.9 per 1000 births

Nongenetic factors can also be involved in fraternal twinning. Weather appears to influence ovulation. Multiple ovulations are more frequent in cold weather; hence, more fraternal twins are conceived during the winter months than in summer months among people of the same racial group. One study showed that when a cold "Texas norther" struck, a rash of twins would be born about nine months later. Also, the chance for fraternal twins increases with age. For a first pregnancy in a woman under twenty, the chance of fraternal twins is about 6 per 1000 in the United States. For a woman in her late thirties, however, the chance goes to about 13 per 1000. Previous pregnancies increase the chance. For a woman in her late thirties who has already had five births, the chance jumps to 21.5 per 1000. Monozygotic twins occur with essentially the same frequency regardless of maternal age or previous pregnancy history.

The proportion of identical to fraternal twins in a population can be calculated from the percentage of twins of the opposite sex. According to chance distribution of the sex chromosomes, about one-half of all fraternal twins will be of opposite sex, so a doubling of the percentage of those of opposite sex will give the percentage of fraternal twins. For instance, about 31 percent of the twins born in the United States are of opposite sex, so about 62 percent of all twins must be fraternal. The rest, about 38 percent, will be identicals.

Conjoined Twins

About 1 out of each 500 pairs of twins born will be joined together (Figure 3–8). They are sometimes called **Siamese twins,** a name given to a pair from Siam exhibited by P. T. Barnum during the last century. There are two kinds. First are those with separate bodies held together at one area, most commonly at the front or side of the chest or abdomen, but it may be at the buttocks or at the top of the head. These can usually be separated by surgery. They may arise when the inner cell mass of a blastocyst separates into two parts that fail to pull apart completely. Or, two areas of embryonic organization may arise and produce two embryos that grow together where they are in close contact. Second are those that share large areas of the body. They may have two heads and chests, but only one set of hips and legs. Or the hips and legs may be double, but they have only one chest and head. These twins usually die before birth or shortly thereafter. They may originate when the inner cell mass separates at only one end, or two centers of embryonic organization may arise at one end after the mass is well along in development, about twelve to thirteen days after conception.

Sometimes the study of conjoined twins causes an undue fear of bearing such children in the minds of some young people. Such twins, however, are far down the line in potential abnormalities. They have a frequency of only about 1 in 50,000 births, while more tragic abnormalities are more common. Down syndrome, or mongolism, for instance, is about eighty times more frequent. In the very rare event that conjoined twins are born, most of them can be separated to give two healthy, normal children. Should the twins be the much less common extensively joined type, they will almost certainly die shortly after birth. A child with Down syndrome, on the other hand, will be more

FIGURE 3–8. The original Siamese twins, found in southeast Asia by P. T. Barnum and exhibited in his circus sideshows.

likely to live and will be greatly retarded mentally, requiring extensive care throughout life. There is no way to avoid all risks during any pregnancy, but the chance of various defects should be kept in proper perspective.

PROBLEMS AND QUESTIONS

1   If parthenogenesis occurred in humans—i.e., an unfertilized egg developed into a child—what would be the sex of the child? Assume that the set of twenty-three chromosomes in the egg were doubled to produce the typical forty-six chromosomes in each cell of the child.

2   Many college and high school textbooks have cited tongue rolling (versus inability to roll the tongue) and hand folding (left thumb on top versus right thumb on top) as examples of simply inherited traits. Studies done in the 1970s, however, suggest that one member of a pair of identical twins may be able to roll his or her tongue while the other cannot. Similarly, one member of a pair of identical twins may fold his or her hands with the right thumb on top, while

the other twin does just the opposite. What may these observations suggest about the inheritance of tongue rolling and hand folding?

3 Place the following events of human development in correct chronological sequence: blastocyst, two-cell stage, fetus, morula, newborn, fertilized egg.

4 Do you know where you can obtain genetic counseling in your state? If you do not know, do you know how to find out? (If your answer to both of these questions is negative, check the references cited in the Answers at the end of the book.)

5 The fact that monozygotic twins are genetically identical is dependent on what biological processes studied in Chapters 1 and 2?

6 The authors mention that increasingly sonograms are used to confirm the presence of multiple embryos in the uterus of a pregnant woman. Sonograms are also done prior to amniocentesis to locate the position of the embryo and the place of attachment of the placenta. A sonogram visualizes the embryo by means of sound waves. Why is a sonogram preferable to an X-ray photograph, which would provide the same information?

REFERENCES

BEACONSFIELD, P., G. BIRDWOOD, and R. BEACONSFIELD. 1980. The placenta. *Scientific American* 243(2):94–102.

CAN GENETIC COUNSELING HELP YOU? (Pamphlet) National Genetics Foundation. (Write: National Genetics Foundation, Inc., 555 West 57th Street, New York, New York 10019.)

FRIEDMANN, T. 1971. Prenatal diagnois of genetic disease. *Scientific American* 225(5):34–42.

FUCHS, F. 1980. Genetic amniocentesis. *Scientific American* 242(6):47–53.

GENETIC COUNSELING. (Booklet) March of Dimes Birth Defects Foundation. (Write: March of Dimes, Box 2000, White Plains, New York 10602.)

GROBSTEIN, C. 1979. External human fertilization. *Scientific American* 240(6):57–67.

HOLDEN, C. 1980. Identical twins reared apart. *Science* 207:1323–28.

———. 1980. Twins reunited. *Science 80* 1(7):55–59.

SHOULD YOU CONSIDER AMNIOCENTESIS? (Pamphlet) National Genetics Foundation. (See address given above.)

STEPHENSON, S. R., and D. D. WEAVER. 1981. Prenatal diagnosis— a compilation of diagnosed conditions. *American Journal of Obstetrics and Gynecology* 141(3):319–343.

Variations in Gene Expression

HEREDITY MAY APPEAR STRANGE AND UN-predictable to one who does not understand its principles. A man with red hair marries a woman with brown hair and all their children have hair like their mother's. When these children marry, however, some of their offspring have the carrot-topped condition of their grandfather. Another couple has two normal children and then a child is born who is mentally defective. They are told that this is an inherited characteristic, yet they cannot understand how this can be when no such children have been born on either side of the family as far back as they can trace. In this chapter we shall show how such cases can be explained through the normal operation of the laws of heredity.

Mendel's Laws of Heredity

In 1866 Gregor Mendel, a monk living in central Europe, published a scientific paper summarizing several years of investigations he had completed on inheritance in garden peas. Mendel set forth two major generalizations or laws that have since been found to apply to many inherited traits in other forms of life. Let us see how these laws operate in human beings.

Genes in Pairs (Alleles)

In Chapter 1 we learned that the chromosomes in a cell can be assembled in homologous pairs, although the members of the sex-chromosome pair in males are of unequal size. Genes also exist in pairs. For each gene at a particular point, or **locus,** on a chromosome, there is a similar gene at the same point on the homologous chromosome.

49

When the two genes of a pair differ somewhat, they are said to be **alleles** of each other. To illustrate, let us assume that a particular gene located on the short arm of chromosome 11 codes the production of an enzyme (tyrosinase) needed for the formation of **melanin.** Melanin is the brown pigment that gives color to the skin, hair, and eyes of most people. An allele of this gene contains a slight variation in the DNA chain and cannot produce the enzyme in a functional form. Many people carry this gene but have normal pigmentation because they also carry its normal allele on their other chromosome 11, and one normal gene can produce all the enzyme needed. When both genes at this locus are defective, however, no functional enzyme is produced and the person is an **albino** (see Figure 4–1). Without melanin the skin is very fair, the hair almost white, and the irises of the eyes a pale blue or pink because of blood showing through. Vision is impaired because, without the light-absorbing melanin in the interior of the eyes, light is reflected about in the interior and damages the retina. In bright light this can be painful, so albinos usually wear dark glasses. The gene that produces the normal enzyme is said to be **dominant;** only one dominant allele is needed for expression. The allele that does not produce the functional enzyme is said to be **recessive** because its effect is not expressed when the dominant allele is also present. Albinism is said to be a recessive characteristic and normal pigmentation is a dominant characteristic.

Terminology and Symbols

When both genes at the same locus on a pair of homologous chromosomes are alike, a person is said to be **homozygous** for that gene. Both may be dominants or both may be recessives. When the two differ, the person is said to be **heterozygous.**

Genotype is a term used to refer to the kind of genes a person has. Three genotypes are possible for dominant-recessive inheritance—two dominants, two recessives, and one of each. **Phenotype** is the term used to refer to the characteristic expressed as a result of the action of the genes. Normal pigmentation is the phenotype of a person who is genotypically either homozygous or heterozygous for the dominant allele for the normal enzyme production. Albinism is the phenotype of a person who is genotypically homozygous for the recessive allele.

Symbols are used for genes in diagramming genetic crosses. A letter symbol is chosen that usually is the first letter of the phenotype that is less common or that deviates from normal. The same letter is used for all alleles at a chromosome locus, lowercase for the recessive allele and capital for the dominant allele. For instance, *a* may be used as the symbol for the gene for albinism, and *A* as the symbol for its normal allele.

FIGURE 4-1. A brother and sister who are albinos lack melanin in their skin, hair, and eyes. They have three brothers and sisters with normal pigmentation. The parents were first cousins, a fact that increases the chance that recessive traits will appear in children.

Most genes that cause deviations from normal are recessive, but some are dominant. For instance, one dominant gene causes a premature closure of the growth areas of the long bones of the arms and legs. The result is shortened arms and legs, which are usually also bowed, but with a normal-sized torso. **Chondrodystrophic dwarfism** is the name of this deviation from normal growth and the capital letter C is used for the gene that causes it. Most people are homozygous for the recessive allele c. This may come as a surprise to some people who have assumed that all recessive genes are less frequently expressed than dominants when, in fact, genes vary in their distribution in populations, and some recessives are much more frequent than their dominant alleles. Sometimes, to avoid confusion, two or more letters are used for a gene symbol. We shall see examples of this as we continue our study.

Gene Segregation and Genetic Ratios

In studying inheritance in garden peas, Mendel concluded that genes occur in pairs in the cells of the body of the plant, but when reproductive cells are produced the reproductive cells contain only one gene of the pair. This illustrates Mendel's principle of gene **segregation** or separation.

Similarly, a person can pass only one-half of his or her genes to any one child. Thus, if a person is heterozygous (A/a) for the gene for albinism, each of his or her gametes will contain only one gene of the pair—either the dominant allele for normal pigmentation (A) or the recessive allele (a) for albinism. Chance governs this segregation of genes in meiosis and determines which allele of a pair any one gamete shall receive. On the average, however, one-half of the gametes will carry the *A* allele and the other one-half will possess the *a* allele. Since each person is heterozygous for many genes, the chance that any two children will receive the same combination of all genes from both parents is so astronomically small as to be considered practically impossible. (Identical twins are excepted because they develop from one zygote.) In fact, even though there are over 4 billion people on the earth, there are so many variant alleles at different gene loci that no two people ever have the same combination. When you look into a mirror, you are seeing a person, the likes of which has never existed before, does not now exist elsewhere, nor will ever exist again in the future.

The way in which allelic genes are segregated when gametes are produced and are then united to form zygotes is shown in Figure 4–2. When both parents are heterozygous for albinism (*A/a*), one-half of the gametes carry the dominant allele and one-half carry the recessive allele. In the chance union of these gametes about one-fourth of the zygotes formed will be homozygous (*a/a*) and will produce albino children. The proportion of all genotypes and phenotypes expected can be represented as follows:

$$A/a \times A/a = A/A + 2\ A/a + a/a \qquad (3 \text{ normal pigment} : 1 \text{ albino})$$

$$\text{(genotype)} \qquad\qquad\qquad \text{(phenotype)}$$

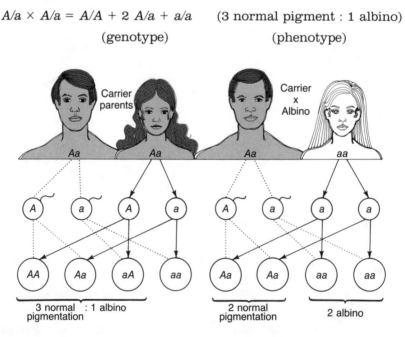

FIGURE 4–2. Distribution of the gene for albinism in the children of two kinds of parents. When both parents are heterozygous, about one-fourth of the children can be expected to be albinos, but when one is heterozygous and the other is an albino (homozygous), the expectation for albino children is one-half.

When one parent is heterozygous and the other is an albino, the expected ratio will be as follows:

$$A/a \times a/a = A/a + a/a \qquad \text{(1 normal pigment : 1 albino)}$$

These calculations give mathematical expectations or probabilities. The actual distribution of genotypes and phenotypes in any one family can deviate considerably. This fact is often misunderstood, as illustrated by the reaction of one set of parents whose first child was an albino. When a genetic counselor explained the mathematical probabilities for albinism, they seemed relieved and said, "We are so glad to know that—now that we have had the albino, we can have three more children who will be normal." Actually, the chance for albinism is just as great at the second birth as it was at the first since chance has no memory and probabilities indicate the chance at each birth regardless of the kind of children at previous births. If we study the children of many families where both parents are heterozygous, however, we would find that about one-fourth are albinos. Some of these families might have four albino children, but these instances will be balanced by other families in which the number with normal pigmentation exceeds the three-fourths of the mathematical ratio.

Reasons for Dominance or Recessiveness

Why are some genes dominant and others recessive? Many dominant genes produce substances necessary for normal life processes. In most cases one such dominant gene can produce all that is needed. Recessive alleles of these genes may not produce the substance, or may produce it in an altered form that does not function properly, as illustrated by the gene for albinism. It should be pointed out, however, that two recessive genes are not always necessary for expression of a recessive characteristic. One alone will be expressed if no dominant allele is present. Males have only one X chromosome, and this carries many genes with no alleles on the Y chromosome. All of these will be expressed (see Chapter 6). Also, sometimes a small piece may break off one chromosome of a pair and be lost to the cell. Any recessive genes at this region on the intact homologous chromosome will be expressed (see Chapter 11).

In some cases a dominant gene causes an abnormality because it produces a substance that interferes with normal reactions. It may inhibit the action of certain enzymes or hormones, slow the absorption of vital materials by the cells, or have some other toxic effect. **Huntington's disease (chorea)** is characterized by a progressive deterioration of mental and motor functions beginning after maturity in most cases. It is caused by a dominant gene that produces a substance that interferes with normal metabolism in the brain. The result is gradual deterioration of brain tissue. Most people are, obviously, homozygous for the recessive allele (h/h).

Multiple Alleles

We have been considering only two possible alternative alleles at a chromosome locus, but there can be more. Twenty alleles have been identified at one of the loci involving proteins in the membranes of red blood cells (see Chapter 8). Any one person can carry no more than two of these alleles, but it is this great potential variation at each gene locus that makes possible the many genetic variations in the human population.

Heterozygous Gene Expression _____

In some cases both genes of an allelic pair are expressed when heterozygous. According to the degree of expression, these fall into three categories.

Intermediate Inheritance (Lack of Dominance)

When both genes of an allelic pair are partially and about equally expressed, inheritance is said to be **intermediate.** We can illustrate with **sickle-cell anemia,** a disease that afflicts about 1 black child out of each 400 born in the United States. Those children homozygous for a certain gene produce a defective hemoglobin **(hemoglobin S)** that forms into rigid chains when the oxygen level of the blood drops below a certain point, as it often does in the veins. This rigidity causes distortion of some of the red blood cells into crescents or sickle shapes (see Figure 4–3). Such cells do not transport oxygen efficiently and are removed from circulation by the spleen, thus causing a severe anemia, as well as other symptoms (see Chapter 10). Heterozygous people generally do not have anemia, but a blood smear will usually reveal a few sickle-shaped cells. An analysis of the hemoglobin shows about equal proportions of hemoglobin S and the normal **hemoglobin A.** Hence, both alleles are functioning and each is producing about one-half of what it would if homozygous. This is intermediate inheritance of hemoglobin. Should the oxygen level of the blood drop considerably, such as might occur when a person goes to a high altitude, many more of the cells may become sickled and anemia may result. These heterozygous people are said to have the **sickle-cell trait** in contrast to the homozygotes who have sickle-cell anemia.

Intermediate expression is symbolized by using superscripts over one basic letter or series of letters. Hb is the symbol for hemoglobin and Hb^A is the symbol for the allele that produces hemoglobin A. Hb^S is the symbol for the allele that produces hemoglobin S. When two persons have the sickle-cell trait (see Figure 4–4), the ratio expected in their children is as follows:

$$Hb^A/Hb^S \times Hb^A/Hb^S = Hb^A/Hb^A + 2\ Hb^A/Hb^S + Hb^S/Hb^S$$

(1 normal : 2 sickle-cell trait : 1 sickle-cell anemia)

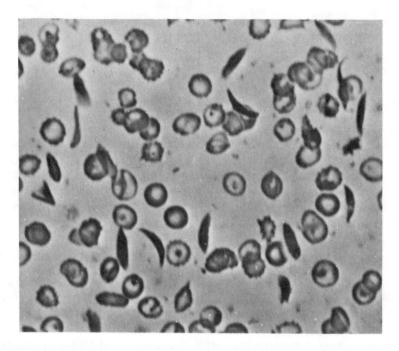

FIGURE 4–3. Blood from a person with sickle-cell anemia showing many of the red blood cells distorted into sickle shapes.

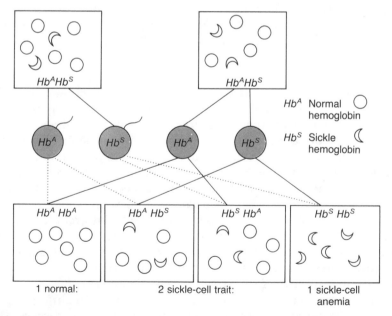

FIGURE 4–4. Inheritance of the gene for sickle-cell anemia. When both parents are heterozygous and have the sickle-cell trait, about one-fourth of their children will have sickle-cell anemia and about one-half will have the sickle-cell trait.

Intermediate inheritance will occur when both genes of an allelic pair produce their substances partially, or when one allele does not make as much of its product as when two are present and the other allele makes no product or makes a nonfunctioning product.

Codominant Inheritance

Sometimes both alleles of a pair act as dominants; that is, they both produce their products in heterozygous individuals. For example, one gene produces what is known as the A antigen in the membranes of red blood cells. An allele produces the B antigen. A person who is heterozygous for these two alleles will have type AB blood; both antigens will be present in the heterozygote. These two alleles, therefore, are considered to be **codominant.**

Heterozygous Influence of Recessive Genes (Carrier Detection)

Even though a gene is considered to be recessive, it still may have some small influence that usually goes unnoticed but sometimes can be detected by special tests. Such detection can be important in cases where the recessive gene causes a serious abnormality when homozygous. Prospective parents are naturally concerned about the possibility of some abnormality in the children they produce. This is especially true when some genetic defect has appeared in close relatives. It would be a great help if they could know whether or not they carry genes for certain defects. A specific example will illustrate how an analysis for heterozygous effects of recessive genes can be of value in genetic counseling.

A child who is homozygous for the recessive gene for **Tay-Sachs disease** is normal at birth but at about five to six months shows a progressive decline in mental ability. This is followed by an impairment of vision, muscular weakness, and a gradual loss of all mental and physical control before a merciful death at about three to four years. Children with this disease lack the dominant gene for production of **hexosaminidase A,** an enzyme needed for processing of sphingolipids (fatty substances). The unprocessed fatty material accumulates in the nerve sheath and slowly chokes off the passage of nerve impulses. Heterozygous persons produce all the enzyme needed for normal fat metabolism, but not as much as those who are homozygous for the dominant gene. This difference can be detected by a simple blood analysis. If both prospective parents find that they are not carriers, they can have children without any fear of this dread disease. If one is a carrier and the other is not, they can still have all normal children, about one-half of whom will be carriers. If both prospective parents are carriers, they will know that the chance for an afflicted child is one-fourth and may decide to forego parenthood because of these high odds. As was brought out in Chapter 3, however, they have an alternative if their moral convictions will allow them to use it. They can start a pregnancy and then have amniocentesis followed by abortion if the fetus cannot produce the enzyme. (See Chapters 9 and 15 for further information on Tay-Sachs disease.)

Cystic fibrosis is a disabling affliction that formerly always caused death by the early teens. We now have treatments that can prolong the lives of persons with cystic fibrosis, but activities are very restricted. In cystic fibrosis, a certain recessive gene is homozygous and the body produces abnormal glycoproteins that interfere with salt metabolism. The sweat of the body is high in sodium chloride (table salt), but more important, the mucus secreted is abnormally viscid and tends to clog passageways in the lungs, liver, and pancreas. The accumulation of mucus in the lungs often leads to pneumonia, and liver function is impaired so that bile production is reduced and fat digestion is not normal. Fibrous growths develop in the pancreas and interfere with the passage of digestive enzymes.

Measurement of the salt content in the sweat was used to try to detect heterozygotes, but it was found that there is some overlapping in quantities when compared to homozygous normal persons. Then, an ingenious experimenter, Barbara Bowman, found that cilia from the siphons of oysters react differently to blood sera from different persons. Serum from one who is homozygous normal will not affect the beating of the cilia, but serum from one with cystic fibrosis will stop the beating and mucus will be secreted. Serum from a heterozygote will also affect the cilia, although not as strongly as that from a person with cystic fibrosis. Ciliated epithelium from the trachea of rabbits was also found to react in this manner. Some feel that these tests for heterozygotes are not completely definitive yet, but they are being used experimentally in many laboratories. Early in 1981 a new test for the detection of individuals heterozygous for the CF gene was announced in the *New England Journal of Medicine.* The test, which requires culturing fibroblasts from suspected heterozygotes, does not appear to be simple enough for mass screening procedures. It will, however, be useful in identifying heterozygotes in families having a history of CF.

Lethal Genes

Genes producing phenotypes that deviate from normal may lower the chance for survival. The degree of reduction of viability is variable. Some genes may cause death in only about 10 percent of those who express the phenotype, others may destroy about one-half, and some may cause death in practically all. The last are known as **lethal genes.**

Time of Lethal Effect

The time at which lethal genes exert their deadly influence varies. Some genes interfere with mitosis of the zygote and life stops before the two-cell stage is reached. Others may cause an abnormal development of the blastocyst and the embryo dies before it reaches the uterus. Others may not be able to produce the enzymes needed for implantation, in which case the embryo is expelled from the uterus. All of these would kill so early that a woman would never know that conception had occurred. A lethal gene that prevents normal formation of the heart, or normal blood production, causes death at about three weeks after fertilization because this is the time when circulating blood becomes vital for continued existence. Others may kill at various times depending on the time their products become vital. One that produces malfunctioning kidneys does not interfere with development during fetal life because the mother's blood absorbs the excretory wastes. At birth, however, the baby becomes dependent upon its kidneys for such removal and dies within a few days. Other lethals causing neonatal deaths involve abnormalities of lungs and shifts in the circulatory system that must channel blood from the heart to the lungs instead of to the umbilical cord.

Usually, we think of lethal genes as those that cause death before or within a few days after birth, but some are late acting and do not exert their deadly effect until later in life. Tay-Sachs disease causes death several years after birth, a type of muscular dystrophy causes death in the teens, and sometimes death can be even later. Huntington's disease usually brings about death at about forty to fifty years of age.

Heterozygous Expression of Lethals

Dominant lethals may appear by mutation but are eliminated in the first generation unless they are of the late-acting type that cause death after it has been possible to have children. Other lethals cause death when they are homozygous, but have a nonlethal phenotypic expression when heterozygous. **Brachydactyly** is the intermediate expression of such a gene. The middle joint of the fingers is so greatly reduced that it appears as though the hand has all thumbs. Statistically, when two persons with such fingers have children, about one-fourth should be normal, about one-half should have brachydactyly, and about one-fourth should be born with no fingers or toes, and with other bone defects that cause death during or shortly after birth. This is the typical 1 : 2 : 1 ratio of intermediate inheritance. It is quite probable that many of the rather rare abnormalities that we have been calling dominant phenotypes are actually the intermediate expression of genes that would be lethal when homozygous. We have had no way to verify this because there has been no record of offspring of two persons with the same abnormality.

Chance of Expression of Lethal Genes

The human body is so complex that literally thousands of genes are necessary for the continuation of life. A mutation of any one of these could make it a lethal gene. It is estimated that each person carries two or three recessive lethals, on the average, but there are so many kinds of lethal genes that your chance of marrying someone with even one lethal matching one of yours is not great. Statistically, should this matching happen, the lethal would be expressed in only one-fourth of your children.

The chance that both parents carry the same lethals increases when they are close relatives. Persons with a common ancestry are more likely to have many genes in common than would nonrelated individuals. As a result, we find more spontaneous abortions, stillbirths, and neonatal deaths in children of relatives than in children from unrelated parents. The exact probabilities for this are discussed in Chapter 15.

Multiple Effects of Single Genes (Pleiotropy)————

Some human traits, such as stature, are known to be regulated by multiple genes at different loci; there are also some single genes that have multiple phenotypic expression, or **pleiotropy.** Such a combination of traits from one basic cause is known as a **syndrome.** Any one gene can do no more than code for one specific polypeptide chain, yet one slight difference in the amino acid sequence of this chain can have far-reaching effects on different parts of the body. We can best illustrate with examples.

Marfan Syndrome

Anyone who receives the dominant gene for **Marfan syndrome** will express a number of traits that may seem to be unrelated, but they all result from one basic defect, the production of an abnormal form of connective tissue (Figure 4–5). This tissue includes bone, cartilage, and tissues that hold various body parts in place. Three categories of defects result.

1   *Skeletal Defects*   The long bones of the body grow longer than they usually do because points of growth do not close as early as they should. Persons with this syndrome are sometimes said to have spider fingers, or **arachnodactyly,** because of the great length of the fingers. The toes are also very long. Because the ligaments and tendons at the joints permit great flexibility, a person with this syndrome can often do unbelievable contortions. Because of unequal growth of the bones of the rib cage, the chest may protrude (pigeon chest), or it may be depressed. Also because of unequal bone

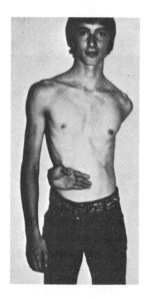
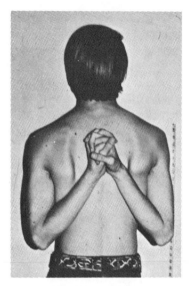
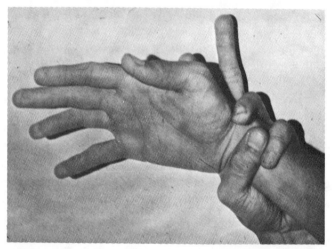

FIGURE 4–5. This young man with Marfan syndrome exhibits the characteristic extreme looseness of joints. He can reach beyond his navel with his arm around his back, can clasp his hands behind him, and the little finger overlaps the thumb when his hand is wrapped around his wrist. (Courtesy Edward Peeples.)

growth, those affected frequently have **scoliosis**, which is the lateral curvature of the spine.

2   *Cardiovascular Defects*   Because of a weakness in the connective tissue within the walls of the aorta, this important blood vessel that leads from the heart may dilate excessively after a heartbeat. The

valves of the heart may also be weak and may allow some regurgitation of blood back into the heart after the beat.

3   *Eye Defects*   Because of the weakness of the connective tissue that holds the lens in place, the lens may be displaced, usually in an upward position. Sometimes it is even possible to see the lower part of the lens through the pupil. The edge of the lens may contact the retina and cause some retinal displacement.

The degree of expression of these defects varies. In some cases there may be only the long limbs and loose jointedness with little inconvenience, while in others there is considerable disability. Abraham Lincoln seems to have had Marfan syndrome, judging by his photographs and descriptions of his long, loose-jointed limbs and very long fingers and toes. Also, there are reports that he had heart trouble, probably due to aortic regurgitation. He had a brother who died in his youth apparently from more extreme manifestations of the syndrome.

Ehlers-Danlos Syndrome

The dominant gene for **Ehlers-Danlos syndrome** results in poorly structured collagen fibers, which are an important part of the skin, cartilages, and walls of the blood vessels. The result is an extreme hyperelasticity of the skin (see Figure 4–6). Because of the great flexibility of the cartilage, there is hypermobility of the joints. There are frequent ruptures of blood vessels under the skin, and often many hematomas that show as blue spots on the skin. Pressure on the skin at

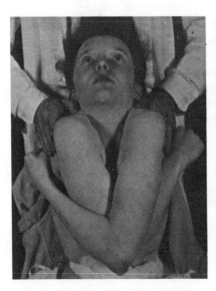

FIGURE 4–6.   Ehlers-Danlos syndrome. The extreme flexibility of the shoulder joints and the great flexibility of the skin and thumb are demonstrated by this young woman.

the elbows and knees causes it to be very loose and furrowed in these regions. Any injuries to the skin tend to heal very slowly. Many individuals with the gene for this syndrome are born prematurely because the amnion is weak and tends to break before full term.

Independent Gene Assortment

Mendel's second principle or law concerns the behavior of two or more genes being considered simultaneously. Mendel concluded that genes affecting different traits behave independently of each other. Today we know that this is true if the genes are on separate pairs of chromosomes. Mendel spoke of the **independent assortment** of different gene pairs.

The principle of independent assortment can be demonstrated in an individual heterozygous for two gene pairs—*A/a B/b*. If the genes *A*(*a*) and *B*(*b*) are on separate pairs of chromosomes, they can align at metaphase of the first meiotic division in either one of two ways:

$$(1) \quad \frac{A}{a}\frac{B}{b} \qquad \text{or} \qquad (2) \quad \frac{A}{a}\frac{b}{B}.$$

When the chromosomes migrate to opposite poles during anaphase of meiosis I, chromosome arrangement (1) above results in gametes containing gene combinations *AB* or *ab*. Chromosome arrangement (2) shown above results in gametes with gene combinations *Ab* or *aB*. Since arrangements (1) and (2) are equally likely to occur, the four gamete types—*AB, Ab, aB,* and *ab*—are produced with equal frequency (i.e., 25 percent of each type). An exception to the principle of independent assortment results when genes being studied are not on separate chromosome pairs, but are located near each other on the *same* chromosome pair. This latter situation, called **linkage,** was not discovered until early in the twentieth century. Closely linked genes do not "obey" Mendel's law of independent assortment.

Assuming independent assortment, what results can we predict from mating two individuals, both with the genotype *A/a B/b*? If the *A* allele is completely dominant to the *a* allele and *B* is completely dominant to *b,* the following logic permits us to answer this question. An *A/a B/b* female is mated to an *A/a B/b* male.

1 If we look only at the *A/a* × *A/a* portion of this mating, we can predict that 3/4 of the offspring will be expected to exhibit the dominant *A* allele and 1/4 the recessive *a*. That is, 1/4 will be *A/A* + 2/4 *A/a* = 3/4 *A*−, while 1/4 will be *a/a*.

2 Similarly, in the *B/b* × *B/b* portion of the mating, we can expect that 3/4 of the offspring will be *B*− in phenotype and 1/4 will be *bb*.

3 Since the probability of two independent events occurring simultaneously is equal to the product of their separate probabilities, we can now predict the overall results of the mating:

$$3/4\ A- \times 3/4\ B- = 9/16\ A-\ B-$$
$$3/4\ A- \times 1/4\ b/b = 3/16\ A-\ b/b$$
$$1/4\ a/a \times 3/4\ B- = 3/16\ a/a\ B-$$
$$1/4\ a/a \times 1/4\ b/b = 1/16\ a/a\ b/b$$

Another way in which these results can be determined is shown in Figure 4–7. In the mating of an *A/a B/b* female with an *A/a B/b* male, each can produce four kinds of gametes—*AB, Ab, aB,* and *ab*—with equal frequency. These gametes can be combined in all possible combinations as shown in the checkerboard diagram (Figure 4–7). Assuming complete dominance of *A* and *B*, one can combine the appropriate genotypes in the checkerboard to give the phenotypic ratio of 9 *A–B–* : 3*A–b/b* : 3 *a/aB–* : 1 *a/a b/b*.

Do you think you could use one of the procedures outlined above to predict the results of the following mating: *A/a B/b C/c* × *A/a B/b C/c?*

Mating: *A/a B/b* female × *A/a B/b* male
Gametes: *AB, Ab, aB, ab* produced by both sexes

Sperm

	AB	Ab	aB	ab
AB	A/A B/B	A/A B/b	A/a B/B	A/a B/b
Ab	A/A B/b	A/A b/b	A/a B/b	A/a b/b
aB	A/a B/B	A/a B/b	a/a B/B	a/a B/b
ab	A/a B/b	A/a b/b	a/a B/B	a/a b/b

Eggs (label at left spanning Ab, aB rows)

FIGURE 4–7. A mating between two doubly heterozygous (dihybrid) individuals, each of whom produces four kinds of gametes with equal frequency. With independent assortment of the *A(a)* and *B(b)* genes and complete dominance of *A* and *B*, the expected phenotypic ratio is 9 *A–B–* : 3 *A–b/b* : 3 *a/a B–* : 1 *a/a b/b*. Can you determine the genotypic ratio from the diagram?

Polygenic Inheritance

Some human characteristics are clearly influenced by heredity but vary greatly from one extreme to the other. This variation in expression is usually an indication of **polygenic inheritance;** that is, genes at more than one chromosome locus are involved in the expression of the characteristic.

Body Stature

It takes no great powers of observation to note that people vary considerably in height. Environment plays a role—diet, disease, and other nongenetic factors during growth can influence eventual stature—but this is not the whole story. Children raised under approximately the same conditions still show considerable variation. Polygenic inheritance is the reason.

To illustrate, let us assume that three pairs of alleles with intermediate relationship are involved in determining potential body height. An individual with genotype *A/A B/B C/C* has the maximum number (6) of genes for tallness and therefore the greatest stature, while an individual with genotype *a/a b/b c/c* has no genes for tall stature. In a population of many individuals, a person could have any one of seven different combinations with 0, 1, 2, 3, 4, 5, or 6 genes for tallness. If a man and woman of intermediate stature, each heterozygous for all three genes (*A/a B/b C/c*), were to marry, they could each produce eight different kinds of gametes—*ABC, ABc, AbC, aBC, Abc, aBc, abC, abc.* Their children could be expected to range more or less continuously,

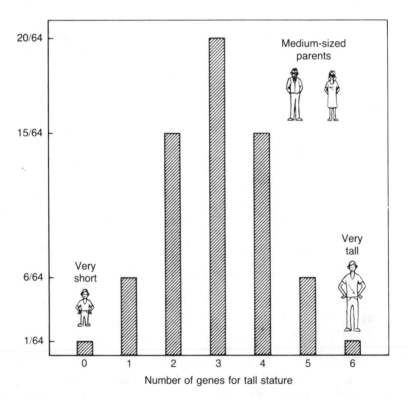

FIGURE 4—8. Distribution of genes for body stature in children of parents of medium stature, assuming that three pairs of alleles are involved.

from those of minimum stature (0 genes for tallness) to those of maximum stature (6 genes for tallness). If we classified a large number of individuals produced by such matings, the distribution would form a bell-shaped curve with the fewest numbers having 0 or 6 stature genes. Most individuals would be clustered toward the middle of the distribution with 2, 3, or 4 genes for tall stature. In other words, because of independent assortment (in meiosis) of the genes affecting stature, medium-sized parents could have a child with any combination from 0 to 6, but the chance would be greatest for those in the mid-range. Figure 4-8 shows that medium-sized parents (each with genotype $A/a\ B/b$ C/c could expect children distributed as follows: 1/64 (0 genes for tallness), 6/64 (1 gene for tallness), 15/64 (2 genes for tallness), 20/64 (3 genes for tallness), etc. Can you verify this distribution using the "checkerboard" procedure shown in Figure 4-7?

Eye Color
The color of the iris of the eyes is often used to illustrate dominant-recessive inheritance, with brown being dominant over blue. One dominant gene does seem to determine that melanin is deposited in the outer layers of the iris, and its recessive allele does not lead to such deposit. It is quite evident, however, that shades of eye color vary considerably. When the dominant allele is present, the eye color may be green, hazel, light brown, medium brown, or very dark brown. The exact color depends upon the amount of melanin present, and this is regulated by polygenes. (It should be mentioned that even those homozygous for blue eyes have melanin in the interior coating of the eye in contrast to albinos, who have no melanin at all.)

When both parents are blue-eyed, all the children would be expected to have blue eyes. There occasionally appears to be an exception. A little investigation will probably reveal that one parent has a greenish tint to the eyes. That parent probably carries the dominant gene for melanin deposit in the iris along with polygenes for a very light deposit. The parent with the true blue eyes, on the other hand, may carry polygenes for heavy melanin deposit, but these cannot be expressed because the gene for no melanin deposit inhibits them. A child can inherit the dominant gene for melanin deposit from the green-eyed parent and the genes for heavy deposit from the blue-eyed parent. The amount of melanin in the eyes is usually correlated with melanin in the skin and hair, yet the genes can be independent. We often see fair blondes with brown eyes and dark brunettes with blue eyes, especially in people with mixed racial ancestry.

Skin Pigmentation
As is true for eye color, a single gene pair can determine whether or not melanin will be deposited in the skin. The gene for albinism prevents melanin deposit, yet it is obvious that great quantitative variation

exists in the amount of melanin in those who have the dominant gene. Polygenes are involved. (Because environment plays a role, we must measure pigmentation in areas of the body that are not usually exposed to the sunlight.) One study reported nine degrees of pigmentation among the people of certain Caribbean Islands where there had been a considerable blending of those of black African ancestry and those of white European ancestry. This would indicate that four pairs of alleles are involved in the pigmentation differences between these two races. The first-generation children of interracial parents would receive four genes for heavy pigmentation and four for light pigmentation. Since all these allelic genes seem to be intermediate, the pigmentation of these children would be about halfway between their parents' pigmentation levels. Children of two persons with such intermediate pigmentation, however, could range all the way from the dark African type to the light European type. Children of one with medium pigmentation and one with light pigmentation could vary from medium to fair, with most about halfway between (Figure 4–9).

Epistasis

The expression of one gene sometimes prevents the expression of another gene or polygenes at different chromosome loci. We have already learned that the recessive gene for blue eyes inhibits the

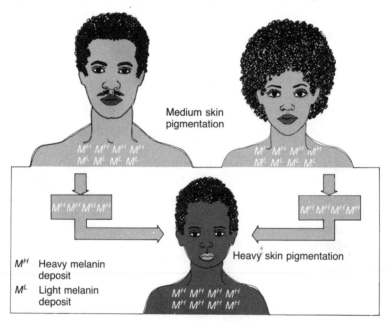

FIGURE 4–9. The chance assortment of polygenes in meiosis of parents with medium skin pigmentation makes it possible for them to have a child with dark pigmentation.

expression of the polygenes influencing the degree of melanin deposit in the iris. This is **epistasis.** Likewise, the gene for albinism is epistatic to the polygenes governing the intensity of melanin in the skin, hair, and eyes. Even though both parents carry genes for very heavy melanin deposits, a child will have no melanin if it also inherits the same allele for albinism from each parent (see Figure 4–10).

A man may be less than four feet tall because he is homozygous for a recessive gene that slows the synthesis of the growth hormone in his pituitary gland, yet he might actually carry polygenes for a very tall stature. The children of such a man and a woman of about average height can be well above average because of genes for tallness from their diminutive father. They would receive the dominant allele for normal pituitary function from their mother. In the same way, a child with genes for high intellectual potential might be mentally retarded because one pair of recessive genes interferes with normal brain metabolism.

The human body is a complex aggregation of organs and systems, and many gene loci are involved in the development and functioning of each of these. A change resulting from the DNA alteration of just one gene in this group can result in the malfunction of an entire organ or system. Hearing, for instance, involves the collaboration of a number of individual parts. One recessive gene can result in improper conduction of impulses by the auditory nerve with resulting deafness, even though all parts of the ear are normal. Another gene might alter the ear bone apparatus that transmits vibrations from the tympanic mem-

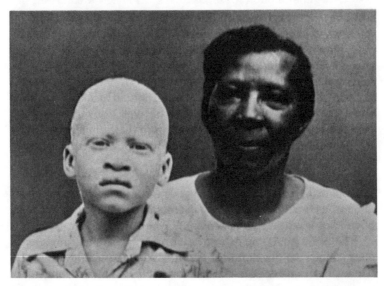

FIGURE 4–10. The gene for albinism is epistatic to the genes for heavy skin pigmentation. This boy is an albino even though both parents have the dark pigmentation characteristic of blacks, as shown by the mother.

brane (eardrum) to the endings of the auditory nerve. Again, deafness results even though the auditory nerve is normal. Hence, so far as hearing is concerned, each of these genes is epistatic to the other. Since both of these genes are recessive, the children of two people with hereditary deafness for different reasons can have normal hearing.

PROBLEMS AND QUESTIONS

1 A normally pigmented man whose mother was an albino married a normally pigmented woman who had had an albino child in a previous marriage. Give the genotypes of all individuals involved and determine the probability that this normally pigmented couple will have an albino child.

2 A normal woman, who does not have cystic fibrosis (CF) and whose parents did not have CF, but whose sister died at age two years of CF, seeks the advice of a genetic counselor. She wants to know the probability that she carries the gene for CF. She also wants to know the probability of her having a CF child of her own. What should the genetic counselor tell her?

3 Most babies born with cystic fibrosis (CF) are the children of normal (non-CF) parents. Most individuals who develop Huntington's disease can be shown to have a parent with Huntington's disease. Are these observations consistent with basic principles of Mendelian genetics? Explain.

4 The general public often confuses sickle-cell anemia and sickle-cell trait. How would you explain the differences between these two inherited conditions?

5 Mendel's law of independent assortment is concerned with matings involving two or more different genes, with each gene being located on a different chromosome pair. Suppose that in an individual with the genotype $A/a\ B/b$, the $A(a)$ and $B(b)$ genes are on two different pairs of chromosomes.
   (a) How many different kinds of gametes can $A/a\ B/b$ produce? What genotypes do these gametes have?
   (b) If $A/a\ B/b$ is mated with $a/a\ b/b$, what is the probability that offspring with the following genotypes will be produced? $A/a\ B/b$; $a/a\ B/b$; $A/A\ b/b$.
   (c) If $A/a\ B/b$ is mated to $A/a\ B/b$, what is the probability that the offspring will express both dominant A and dominant B? will express dominant A and recessive b? will express both recessive a and recessive b?

6 The biochemical pathways leading to the formation of melanin pigment involve several steps regulated by different genes and their enzymes. Absence of the enzyme tyrosinase, due to being homozygous recessive for a particular gene, produces "classical" albinism.

Other forms of albinism, regulated by other recessive genes, are also known to occur. This sets the stage for explaining the results of different matings in terms of epistatic gene interactions. The following matings and their results have all been noted:

(a) normal × normal ———→ all normal
(b) normal × normal ———→ 3 normal : 1 albino
(c) normal × albino ———→ all normal
(d) normal × albino ———→ 1 normal : 1 albino
(e) albino × albino ———→ all albino
(f) albino × albino ———→ all normal (!)
(g) albino × albino ———→ 3 albino : 1 normal
(h) normal × normal ———→ 9/16 normal : 7/16 albino

Can you offer an explanation for each of these matings, suggesting the possible genotypes of parents and offspring?

REFERENCES

BRESLOW, J. L., J. McPHERSON, and J. EPSTEIN. 1981. Distinguishing homozygous and heterozygous cystic fibrosis fibroblasts from normal cells by differences in sodium transport. *The New England Journal of Medicine* 304(1):1–5.

HARTL, D. L. 1977. *Our Uncertain Heritage: Genetics and Human Diversity*. J. B. Lippincott Co. Philadelphia.

HENDIN, D., and J. MARKS. 1978. *The Genetic Connection*. William Morrow and Co., Inc. New York.

MENDEL, G. 1865. Experiments in plant hybridization. A translation of Mendel's paper appears in J. A. Peters. 1959. *Classic Papers in Genetics*. Prentice-Hall, Inc. Englewood Cliffs, N.J.

MILUNSKY, A. 1977. *Know Your Genes*. Houghton Mifflin Co. Boston.

NAGLE, J. J. 1979. *Heredity and Human Affairs*, 2d ed. C. V. Mosby Co. St. Louis.

THOMPSON, J. S., and M. W. THOMPSON. 1980. *Genetics in Medicine*, 3d ed. W. B. Saunders Co. Philadelphia.

The Determination of Sex

THE DIFFERENCES BETWEEN THE SEXES ARE more extensive than is often realized. In addition to the obvious differences in the organs related directly to reproduction and nursing, nearly all parts of the body show some sexual distinctions. With only a bit of muscle, skin, bone, or blood to go by, a technician can usually correctly identify the sex of the person from which these parts were taken. Whatever forces determine which sex an embryo shall become have an influence that extends to all parts of the body. Two forces are involved—chromosomes and hormones. Let us examine the role of each.

Sex Chromosomes

All persons carry genes for the development of both sexes. You may be a man with heavy musculature and a hairy chest, yet you have the genes required to produce a female. Likewise, even though you are a perfectly normal woman, you carry genes that, if expressed, would enable you to express many masculine characteristics. The determination of sex, therefore, must involve some sort of trigger that allows the genes for one sex to be fully expressed while suppressing those genes for the opposite sex. Sex chromosomes provide the trigger.

X Chromosomes and Y Chromosomes

As was brought out in Chapter 1, males and females differ in one of the twenty-three pairs of chromosomes. In the male, one of these pairs consists of a medium length X chromosome and a much shorter Y chromosome. The Y chromosome is one of the shortest chromosomes in the cell, only about one-fifth the length of the X chromosome. Females have two X chromosomes and no Y chromosome. The X and Y chromosomes are the **sex chromosomes.**

Segregation in Meiosis

During spermatogenesis the X and Y chromosomes pair and then segregate to opposite daughter cells, so about half of the sperm receive the X and the other half receive the Y. Eggs all receive a single X. Therefore, when an X sperm fertilizes an egg, the zygote will have the XX combination and will become a female. A Y-sperm fertilization, on the other hand, will give the XY combination and a male results (see Figure 5–1). The question is, "How can this chromosome difference trigger the expression of so many genes that are necessary to determine sexual characteristics?" We can get a partial answer by studying the effect of abnormalities that sometimes occur in the distribution of the sex chromosomes.

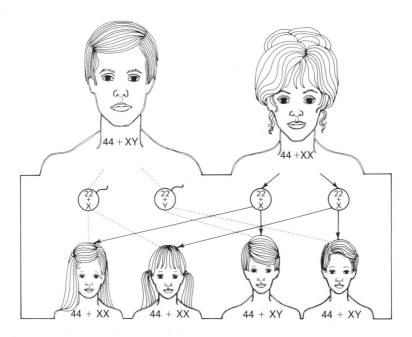

FIGURE 5–1. Method of sex determination. All eggs carry an X chromosome, but sperm are of two types, one carrying an X and the other carrying a Y, and the sex of a child depends upon which type of sperm fertilizes the egg.

Abnormal Sex Chromosome Distribution

Sometimes the two X chromosomes of a female adhere during anaphase I of meiosis and both go to one oocyte and none to the other (see Figure 5–2). As a result of this nondisjunction, an egg may be produced with two Xs, or it is equally probable that an egg will have no Xs. When such eggs are fertilized with normal sperm, various abnormal sex-chromosome combinations result.

1   *XX Egg Fertilized by an X sperm*   The result is a zygote with three Xs, termed *trisomy-X*, which will develop into a female with the **triple-X syndrome.** Many with this syndrome are females who are normal in appearance and may be fully fertile; in fact, they may not even know that they have an extra X unless it is revealed by karyotyping for other reasons. Many others, however, are sterile and have amenorrhea or other menstrual disturbances. The chance for mental retardation is about twice that for normal XX females. About 1 of every 1200 female live births has trisomy-X.

2   *XX Egg Fertilized by a Y sperm*   The XXY zygote that results becomes a male with **Klinefelter syndrome.** During childhood, those with this syndrome appear to be normal males, but at puberty abnormalities become apparent. Their testes do not fully develop and remain only about half the size of those of normal males. Klinefelter males tend to grow taller than average and muscular devel-

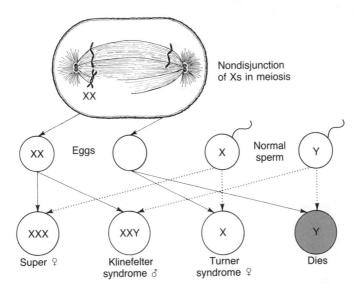

FIGURE 5–2.   Nondisjunction of the X chromosomes in oogenesis results in some eggs with XX and some with no X. When such eggs are fertilized by normal sperm, several types of abnormalities of sex result.

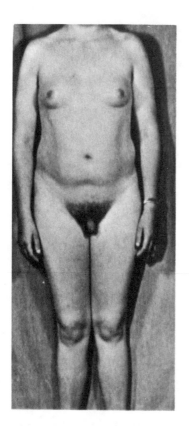 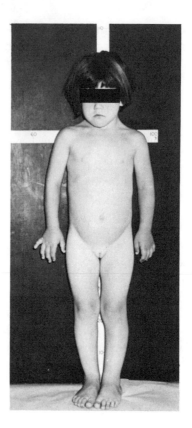

FIGURE 5—3. The XXY sex-chromosome combination causes Klinefelter syndrome, shown at left. The person is a male, but with some breast development, feminized musculature, and reduced size of the sex organs. The 45, X combination results in Turner syndrome, shown at right. The child is a female, but her sex organs will never mature. (Left courtesy Povl Riis; right courtesy Dr. Stella B. Kontras, Children's Hospital, Columbus, Ohio.)

opment is somewhat feminine in nature (see Figure 5–3). They may have some breast development (gynecomastia), and their voices will be higher pitched than normal for males. They can have sexual intercourse but are sterile because spermatogenesis is very abnormal and no fertile sperm are produced. The chance for mental retardation is increased—about one percent of the patients in mental institutions have this syndrome, yet it appears in only about 1 in each 400 live male births. Figure 5–4 is a karyotype of a male with Klinefelter syndrome.

3 *No-X Egg Fertilized by an X Sperm* The resulting XO zygote produces a female with **Turner syndrome.** Before puberty, children with this syndrome may appear to be normal, although they will be

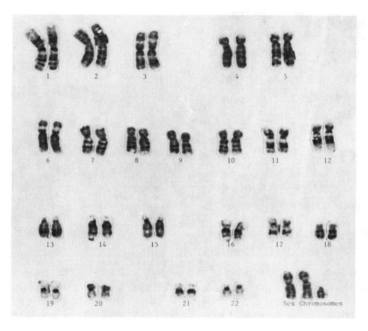

FIGURE 5—4. A karyotype consisting of G-banded chromosomes obtained from a male having Klinefelter syndrome, 47, XXY. (Courtesy Catherine G. Palmer, Indiana University School of Medicine.)

somewhat shorter and more chunky in body build than other girls their age. The most readily distinguished characteristic at birth is a thick fold of skin (Figure 5—5) on either side of the neck ("webbing of the neck"). This makes the neck appear very wide from the front or back. At puberty, the female organs fail to mature and remain in the childlike state. There is no menstruation or breast enlargement. The ovaries do not produce eggs. About 1 female of each 2500 born alive has this syndrome, but the number of spontaneously aborted fetuses with it is quite high. Some estimates indicate that the live-born babies with Turner syndrome represent no more than about 5 percent of the total XO zygotes produced. This would mean that about 1 of each 250 conceptions would be XO. It would seem that the number of XO zygotes produced would about equal the number of XXY zygotes if nondisjunction were always the cause of these abnormalities. Eggs without an X can be produced in another way, however. The sex chromosomes are among the last to separate and move to the poles during meiosis. Sometimes an X may lag so far behind the other chromosomes that the nuclear membrane is formed before the X reaches the pole. Left out in the cytoplasm, this X disintegrates and an egg without an X is formed. Such an egg

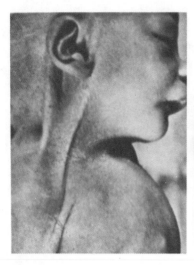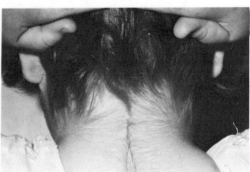

FIGURE 5–5. Characteristics of Turner syndrome. At birth there is a fold of skin on the neck. The photograph on the right (Courtesy, Dr. Stella B. Kontras, Children's Hospital, Columbus, Ohio) shows the characteristic low, irregular hairline on the nape of the neck of the 45, X child shown in Figure 5–3.

fertilized by an X sperm will also produce an XO zygote. Figure 5–6 is a karyotype of a female with Turner syndrome.

4 *No-X Egg Fertilized by a Y Sperm* Zygotes with this combination cannot survive because the X carries genes vital for existence. Every person must have at least one X; hence, YO individuals have never been encountered.

Other Abnormal Distributions

Nondisjunction of the X and Y can occur in the first division of spermatogenesis and will give some sperm with both X and Y and some with neither. An XY sperm fertilizing a normal X egg will give the XXY combination and Klinefelter syndrome results. A sperm with no sex chromosome will give an XO zygote and Turner syndrome occurs.

Nondisjunction can also take place in the second division of meiosis. In the female the results would be the same as if it had occurred at the first meiosis—XX and O eggs would be produced. In the male, however, nondisjunction in the Y-carrying secondary spermatocyte could result in two Ys being in one sperm and none in the other. A YY sperm which fertilized a normal X egg would produce an XYY zygote. This combination was first reported by A. A. Sandberg et al., in 1961. The man having the extra Y was white, 44 years old, of average intelli-

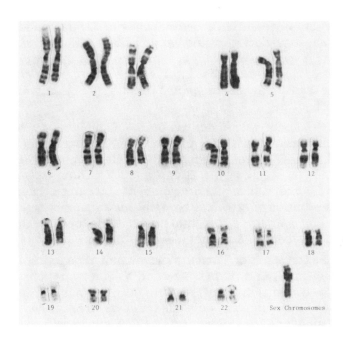

FIGURE 5–6. A karyotype consisting of G-banded chromosomes obtained from a female having Turner syndrome, 45, X. (Courtesy Catherine G. Palmer, Indiana University School of Medicine.)

gence, and without physical defects. He had fathered eight children, including one with Down syndrome (see p. 176). In 1965 a study of violent criminals in Scotland by Patricia Jacobs showed that about four percent of them had XYY karyotypes. These men were above average in height and below average in IQ. Individuals who are XYY are sometimes said to have Jacobs karyotype or **XYY syndrome.** Since 1965, however, routine chromosome analyses have shown that many men leading normal lives are also XYY. In fact, XYY was found to be a common chromosome aberration, occurring in about 1 of each 1000 live male births. This is still a lower frequency than that found in the study of criminals, however, and the concept that there might be some connection between the XYY combination and criminal behavior has persisted.

In an effort to get some additional information, two Harvard professors, Stanley Walzer and Park Gerald, began to identify sex-chromosome abnormalities of each baby born in a Boston hospital and planned to follow up any cases of abnormalities to see if behavior in later life was influenced. A serious objection was raised by some people, however, on the grounds that parents, knowing for instance that their son had an XYY combination, might treat him differently and this might actually result in aberrant behavior. Or the son might learn of his chromosome deviation and would behave differently because he would think it was expected of him. As a result of pressure placed upon them, Walzer and Gerald discontinued the study, even though the majority of the Harvard faculty voted that its benefits outweighed any possible difficulties.

Still other abnormal sex-chromosome combinations have been found. Combinations of XXXY and even XXXXY have been noted and are associated with extreme forms of Klinefelter syndrome and mental retardation in most cases. Also, XXXX and XXXXX have been found in females with abnormal sexual development and extreme mental retardation. The addition of extra Xs above the normal number increases the chance for and the degree of mental retardation. Just how the cells get these high numbers of Xs is not clear at this time.

A partial expression of Turner syndrome may be found in females with one normal X and part of a second X. Sometimes chromosomes break and lose a portion not attached to the centromere. The degree of expression of the syndrome depends on how much of the X has been lost. In particular, if the short arm of the X chromosome is missing, the usual Turner syndrome phenotype results. Partial expression of Klinefelter syndrome has also been found when there has been an intact X and Y plus part of another X.

Sex-Chromosome Mosaics

Occasionally a girl expresses all the characteristics of Turner syndrome, but a karyotype made from blood lymphocytes reveals two Xs. In other cases a karyotype may show only one X, yet there are no symptoms of the syndrome. Studies of the chromosomes of cells from other parts of the body, however, may show that these females are mosaics—some cells are XO and some are XX. This can happen when one of the Xs is lost in a cell of the early embryo. We have already mentioned that the X is the last chromosome to reach the poles in meiosis and sometimes there can be a lag that causes it to be left out of the nucleus. Left out in the cytoplasm, it will disintegrate and be lost to the cell. The same thing can happen in mitosis, and when it occurs in a cell of the very early embryo, an XX/XO mosaic is formed. The phenotype of a mosaic patient depends on the time of the mitotic error in chromosome distribution and the proportion of abnormal cell lines in different organs of the body. Likewise, XXY/XY mosaics can be produced through the loss of an X in cells descending from an XXY zygote. Such mosaic individuals may be normal or have Klinefelter syndrome, depending upon how early in development the mosaicism occurred and the proportion of abnormal cell lines in different organs of the individual.

Sex-Chromosome Recognition in Interphase Cells

Through hormone therapy, surgery, and other methods, it is possible to improve the physical and psychological condition of those with sex-chromosome abnormalities. They can be made to fit better into the pattern of maleness and femaleness that we consider normal. The earlier in life that such treatment is started, the better the results. Unfortunately, many physicians fail to recognize the abnormalities at birth. As a result, treatment is delayed until the abnormalities become apparent, which may not be until adolescence. Karyotyping of all newborns would be an almost impossible task because of the long, painstaking work required for each karyotype. There are relatively simple ways, however, to determine the sex-chromosome complement.

Barr Bodies (Sex-Chromatin Bodies)

In 1949 Murray Barr found that nuclei of interphase nerve cells of female cats have a dark-staining body not found in nuclei of male cells. Human interphase cells were also found to have this sexual distinction (see Figure 5–7). Cells scraped from inside the mouth, when stained with dyes specific for DNA, show a half-moon-shaped body lying against the nuclear membrane when they are from a female. This became known as the **Barr body,** or **sex-chromatin body.** No such body is found in male cells. Other cells also show this Barr body if they are from a female.

Cells from a person with Turner syndrome will have no Barr bodies, while those from a person with Klinefelter syndrome will have one in each cell. Triple-X females will have two Barr bodies in each cell. The number of Barr bodies is one fewer than the number of X chromo-

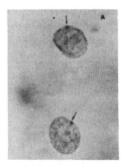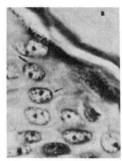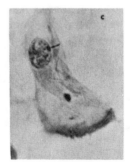

FIGURE 5–7. Female cells showing Barr bodies against the nuclear membrane. (A) Epithelium cells from the lining of the mouth. (B) Cells in a section of the skin. (C) Cells from the lining of the vagina. (Photos courtesy Murray Barr.)

somes. How can we explain this? A Barr body is a tightly coiled X chromosome lying against the nuclear membrane. Gene balance in the cells is very important. In males, the genes on the one X have a certain ratio to the double set of genes on the autosomes. Females have twice this number on their two Xs and would have a very different ratio to the genes on the autosomes were it not for the inactivation of one X. Genes in the tightly coiled X of the Barr body cannot function. Hence, the gene ratio is the same for both sexes. All XX women are mosaics, with one X functioning in some cells and the other X in other cells. Studies by Mary Lyon indicate that chance determines which X is inactivated in a given cell, but once this is determined, all the descendants of the cell will have the same X inactivated. The determination takes place early in embryonic life, probably before implantation. As a result, rather large islands of tissue may express the genes on one X, while other islands will express the genes on the other X. In the next chapter we shall learn how this influences the expression of X-linked genes in females.

If only one X functions in any one cell, then why are females with XO not normal? Also, why does XXY not result in a normal male? The second X in both instances must have some function. It may fulfill this function in the very early embryo before Barr bodies are formed. Moreover, some cytological studies have found what appears to be a part of the X extending out from the Barr body; this part may uncoil during interphase and the genes on it could function. This portion could contain genes related to triggering sex characteristics. In fact, some recent studies show that several genes on the short arm of the X chromosome are *not* inactivated, suggesting that perhaps only the long arm of the X is involved in the formation of sex chromatin.

Leucocyte Drumsticks

The white blood cells known as granulocytic polymorphonuclear leukocytes also show distinctions correlated with the number of X chromosomes. These cells have two or three lobes to the nucleus, and normal females also have a drumstick-shaped body extending out from one of the lobes. The number of drumsticks is one fewer than the number of X chromosomes, so a drumstick must represent a coiled X. Since the drumsticks are often folded over the top of the nucleus, however, and will not be seen in all of these cells, counting drumsticks is not a good method to determine the number of Xs.

Fluorescent Y Chromosomes

Barr body studies reveal abnormal distribution of X chromosomes but do not show the number of Y chromosomes. The Y can be revealed, however, by staining cells with acridine dyes and viewing them under the ultraviolet microscope (Figure 5–8). The Y will stand out as a

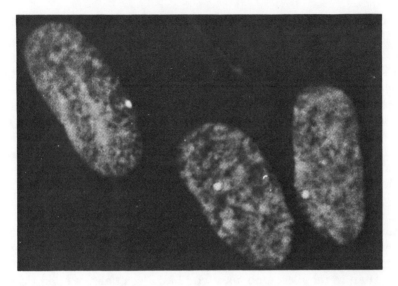

FIGURE 5–8.   The Y chromosome can be identified as a bright spot in these Wharton jelly cells from the umbilical cord. The cells have been stained with quinacrine mustard and photographed under an ultraviolet light microscope.

bright, glowing body in interphase cells. The Y carries very few genes and it is the nongenic portion of chromosomes that glows most brightly (see Chapter 1).

Identification of Sex Chromosomes in Newborns

Many hospitals now check the sex chromatin of all babies born. The placenta, with its attached umbilical cord and membranes, is sent to the laboratory where a bit of amnion is removed, spread on a microscope slide, stained, and the number of sex-chromatin bodies is determined. If the phenotypic sex has been reported as female and there are no Barr bodies, then Turner syndrome is suspected. If the baby is reported as a male and there is a Barr body, then Kleinfelter syndrome is suspected. Two Barr bodies, of course, would indicate triple-X females or XXXY males. The umbilical cord contains large Wharton jelly cells that swell when exposed to the air. This creates a pressure that seals the broken end of the cord and prevents excessive bleeding in animals where there is no tying and cutting of the cord. When the large end of the cord is pressed against a microscope slide, many of these cells adhere and, when stained and viewed under the ultraviolet microscope, the number of Y chromosomes is easily determined. Early identification of sex-chromosome irregularities permits the early initiation of hormone treatment and other measures that can cause the baby to grow into as normal an individual as possible.

Identification in Female Athletes

In all international women's athletic competitions, it is now customary to test for the presence of Y chromosome in the contestants. Some males with Klinefelter syndrome might have breast development and body structure that would enable them to pass as females. Also, some XY males may inherit a gene for testicular feminization (androgen insensitivity) and become so feminized that they themselves may not know that they have internal testes rather than ovaries. (This is described more fully later in this chapter.) The testing program was initiated when it was found that a famous Polish athlete competing in women's events had a Y chromosome and testicular feminization. The analysis may be made by a fluorescent study of hair follicles or from cells scraped from inside the mouth. Some have claimed that this is unfair discrimination against those who are basically female, but have a Y, and that other tests of femaleness should be used. Alternative tests include analysis of protein density of the muscles, which is higher in males than in females. One man who had a sex change operation and wanted to play as a woman in women's tennis tournaments claimed discrimination because the presence of the Y chromosome indicated the individual was biologically a male.

Sex Hormones

Hormones play an important role in the expression of characteristics related to sex. A boy castrated before puberty never develops the beard, voice, musculature, nor desire for relations with women that characterize those with normal male hormone secretions. An overproduction of male hormones early in life results in precocious development of male characteristics. The relationship between female hormones and female characteristics has also been noted. If hormones play such an important role, how are they related to the sex chromosomes?

Early Sex Differentiation

Early embryos show no differences according to sex. Sex organs begin to develop but they appear the same for both sexes. Beginning about the seventh week, however, some influence causes the beginning of differentiation. If the fetus is to become a male, the genital tubercle begins to enlarge to make a penis, the vagina and uterus remain rudimentary, and the outer lips form a scrotum to receive the testes. The gonads of the early embryo have both testicular and ovarian tissues, but one grows and begins functioning while the other regresses. A specific gene on the Y chromosome produces a protein substance that triggers the hormone output of the testicular tissue and this results in male development. This protein contains an **H-Y antigen** and can be identified by immunological tests (see Chapter 7). Because this gene is present in sperm, a possible future method of birth control might involve vaccination of women with this antigen. Antibodies produced in response to this antigen would inactivate any sperm deposited in the vagina, rendering them incapable of fertilization. A double dose of this H-Y gene, as is present in XYY males, results in a high concentration of the antigen. A few cases have been found where an XX embryo becomes a male because a portion of the Y containing the H-Y gene has become attached to an X. The basic human body structure appears to be female; to be male something has to be added. The androgenic hormones, primarily testosterone, bring about this addition.

Stages of Sex Differentiation

The following five events are involved in establishing the sex of a person.

1 *Gene and Chromosome Differentiation*   Cellular sex is determined at conception according to which one of the two kinds of sperm fertilizes the egg. Normally the Y chromosome, with its gene for the H-Y antigen, gives a zygote with male potential.

2 *Gonad Differentiation*   The H-Y gene on the Y chromosome produces the substance that stimulates testicular tissue of the primordial gonads to enlarge and begin the output of androgenic hormones. Without this gene the ovarian portion develops and starts estrogen production. Two Xs are required, however, for normal development of the ovaries.

3 *Anatomical Differentiation*   The hormone output causes the organs characteristic of one sex to enlarge and become functional, while those of the opposite sex remain rudimentary.

4 *Brain Differentiation*   The **hypothalamus** of the brain must be conditioned by early hormone output if all aspects of sex are to be normal. The production of testosterone as early as six weeks conditions the hypothalamus to respond properly to testosterone in later life. Even if extra testosterone is given after birth, sexual development will not be completely normal for a male without this early conditioning. It is possible that some of the cases of anatomical males who feel that they are psychologically female are a result of the failure of this conditioning before birth. A powerful drug or other substance used by mothers of these males in early pregnancy may have interfered with the output of testosterone or its influence on the brain. A number of such males have had surgery to remove sexual organs they felt did not rightly belong to them, and by taking female estrogen have achieved a degree of sex reversal.

5 *Social Differentiation*   After birth, many patterns of behavior and reactions are conditioned by social contacts. A boy soon learns that

certain reactions are manly and should be exhibited, while others are feminine and should be restrained. Girls learn that some reactions are not characteristic of females and others are expected of females. Such imposed patterns are becoming less pronounced today and may not play such an important role in the future.

Androgen Insensitivity (Testicular Feminization)

Sometimes women with problems of painful intercourse or inability to conceive are found to be XY. Externally they may appear to be normal females, but they do not have a uterus, Fallopian tubes, or ovaries, and the vagina may be reduced in size. They actually have testes, with the connecting male tubules, but these remain within the body cavity and do not produce sperm. This condition is known as **androgen insensitivity,** or **testicular feminization.** The cells are not able to respond to the testosterone secreted by the testes. An X-linked recessive gene in mice causes this condition and pedigree studies indicate that it is inherited in the same way in humans, although some studies suggest that it could be caused by an autosomal dominant gene. In the early embryo the Y chromosome stimulates the development of testicular tissue and testosterone is produced, but the cells of the body, particularly those of the hypothalamus, are unable to convert this hormone properly into its active by-products. As a result, the embryo is not conditioned to become a male. Males always produce some estrogen, and the cells can respond to this hormone. Hence, the physical development in androgen insensitivity is female, but insufficient estrogen is present to stimulate development of the uterus, although the vagina develops to some degree.

A high proportion of females in successive generations may be the result of the action of the gene for androgen insensitivity. This gene can be carried by women without any indication of its presence, except for the statistically high number of girls among their children. Figure 5–9 shows a pedigree with an excess of females because of this gene.

Hermaphroditism

In very rare cases a child is born with both sets of sex organs and physical characteristics intermediate between the two sexes. Such persons are known as **hermaphrodites.** Hermaphroditism is normal in some lower animals, such as earthworms, where each animal has the sex organs of both sexes and serves as both male and female in reproduction. In higher animals, however, the sexes are usually separate, except for rare accidents that lead to the hermaphroditic condition. About 34 percent of human hermaphrodites have an ovary on one side and a testis on the other. About 20 percent have ovotestes—that is, part

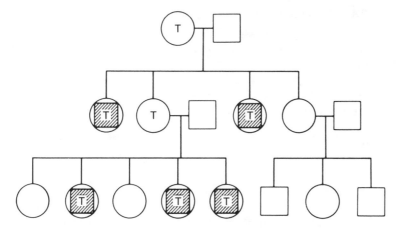

FIGURE 5–9. Pedigree showing inheritance of the X-linked gene for testicular feminization (androgen insensitivity). The gene can be carried by normal females, but when it is passed to an XY child it causes development of a female phenotype. The circles enclosing squares indicate such individuals. The inheritance pattern is characteristic of an X-linked recessive gene.

testis and part ovary—on both sides. The remaining 46 percent have an ovotestis on one side and an ovary or testis on the other.

Since the amount of tissue of one kind or the other varies, there are differences in the amount of hormone output with consequent variations in the degree of intersexuality. Today physicians usually decide which sex seems to be predominant and remove the tissue of the other gonad shortly after birth so development can proceed toward one sex or the other. If the characteristics are about equal the parents might be asked which sex they prefer and the gonadal tissue of the opposite sex is removed. Sexual development can never be fully normal in such cases, but with plastic surgery much can be done toward making the genital organs conform to the sex chosen.

Several theories have been proposed to explain mammalian hermaphroditism. One holds that there is a double ovulation and fertilization. Then the two zygotes, or young embryos, may fuse to form one while still in the Fallopian tube. One of these could be XX and the other could be XY. This could result in bilateral sex tissue with an ovary developing on the XX side and a testis on the XY side. Another theory holds that a single egg nucleus divides before fertilization and a Y sperm fertilizes one nucleus, an X sperm the other. Also, a lagging Y chromosome in an early XXY embryo could result in a line of cells with XX, which would produce an ovary or part of an ovary. Still another theory holds that although the embryo is XY, the ovarian tissue starts development before the male-determining genes on the Y take over and an ovotestis results.

Female Pseudohermaphroditism

The cortex of the adrenal glands produces both male and female sex hormones in both sexes, but these are usually overshadowed by the output of hormones from the testes or ovaries. When there is an excessive development of the adrenal cortex in females, however, the amount of male hormone causes masculinization to a certain degree and the person may be a **pseudohermaphrodite.** Female hormone from an overgrown cortex in the male does not seem to be sufficient to cause hermaphroditism.

Variations in the Sex Ratio

Records in the United States show that about 105 males are born for each 100 females. How can we explain this when meiosis should give an equal number of X and Y sperm? Several possible explanations have been suggested. During meiosis and sperm formation, cells with the larger X may not mature and form sperm as efficiently as those with a Y, so more Y sperm are produced. Also, even though the two kinds of sperm are produced in equal numbers, the Y sperm may have some small advantage in penetrating the barrier of follicle cells around the egg. Since the Y is smaller than the X, the Y sperm could be slightly lighter in weight and smaller in diameter. This could conceivably make a difference in the tight squeeze between the follicle cells. Still another possibility is that the sex ratio is equal at conception but that there are more prenatal female deaths. We know that more than 20 percent of all conceptions fail to produce live-born babies. Could it be that more females are among these prenatal deaths? A tabulation of spontaneously aborted embryos in the great collection at the Carnegie Institution in Washington, D.C. indicates that the reverse is true. Of the embryos old enough to show sexual distinctions (older than eight weeks), 3003 were males and 2784 were females, a ratio of 107.9 : 100. When we discovered ways of identifying cellular sex, however, a greater number of females were found in those younger than eight weeks. K. Mikamo found only 42 males out of 108 such young embryos, a ratio of about 63 : 100. Other studies have supported this finding. Hence, it is possible that conceptions are approximately equal, but that there is a greater female loss early and a greater male loss later in the gestation period.

Immunogenetic investigations suggest some possible explanations for these variations in embryonic survival. The H-Y antigen, from the gene on the Y chromosome, exists in all normal males, but in no normal females. Hence, a woman carrying a male fetus may become sensitized to this foreign antigen. This would be similar to sensitization to the Rh-positive blood antigen (see Chapter 8). Antibodies she produces could enter the male fetus in sufficient quantities to react with the fetal protein and result in death followed by spontaneous abortion. Since female embryos would never be affected in this way, this could account for the higher proportion of males in the spontaneously aborted embryos older than eight weeks. It could also explain the fact that the male ratio declines with each succeeding birth. First births average a ratio of 106.6 : 100, but by the seventh birth the proportion of males has declined to about 104.5 : 100. The antibody production would be accentuated with each male birth.

Immunological reactions could also account for the possible higher death rate of females in very young embryos. The Xg locus on the X chromosome can be occupied by one allele, Xg^{a+}, which produces an antigen, also designated as Xg^{a+}. An alternate allele, Xg^{a-}, does not produce this antigen. A male embryo receives his single X from his mother and would always be compatible with her; he could never have the Xg^{a+} antigen unless she also had it. A female embryo, on the other hand, receives one of her two Xs from her father and could receive the gene for the antigen from him when the mother is homozygous Xg^{a-}. This might be a type of antigen that could cause early sensitization of the mother and early rejection of an incompatible female embryo.

The male : female sex ratio at birth varies among different population groups. It is about 105.6 : 100 for American whites and about 103.3 : 100 for American blacks. The proportion of males is high in Arab countries, Korea, and the Philippines, but low in some countries, such as Pakistan and Jamaica. In Iran, for instance, it is 109 : 100 and in Bangladesh it is 103.5 : 100. Family size could account for some of these differences. The male ratio declines with each succeeding birth so the overall sex ratio would be expected to be higher in areas where small families are the rule and lower in areas where large families are prevalent. It would also be higher in populations where male children are the most desired on the average. If sons are the most abundant among first births in a family, conceptions may be halted. If the first births are mostly daughters, however, the conceptions may continue until a certain number of sons is obtained. Such a practice would result in a high male to female ratio.

The greater survival value of the female also is manifest in multiple births. The male : female ratio for twins in the United States is 103.5 : 100, for triplets it is about 98 : 100, and for quadruplets it is 70 : 100. Males seem to have a reduced chance for survival under stress and the competition among embryos in the same uterus could explain the greater death rate of the males.

Regardless of how the excess of males at birth is explained we know that the ratio declines each year after birth. In the United States females outlive males by seven or eight years on the average. Men may have a higher protein content of their muscles and this gives them a greater physical strength, but women are favored in those qualities that make for survival. By age fifty the higher rate of male deaths has reduced the male : female ratio to about 85 : 100, and at age 85 it is only 50 : 100. Among those who live to the century mark the ratio drops to about 20 :100.

How can we explain the female survival advantage? For one thing, those hormones that give a man a greater size and physical strength also give him a higher rate of metabolism, a faster heartbeat, and a higher blood pressure, on the average. Perhaps a man just "burns out" sooner. Preadolescent children, however, do not have these physiological differences, yet males still have a higher death rate. The answer may be found in the sex-chromosome difference. Most harmful genes are recessive. X-linked genes are haploid in males and all recessive genes are expressed. Some of these genes may be lethal, exerting their effect at different times of life. Others would reduce the chance of survival under stress. Females are diploid for their X-linked genes; therefore, most harmful recessives are covered by a dominant normal allele.

Heterozygote superiority also could be involved. A person heterozygous for certain alleles may have a greater chance for survival than one who is homozygous for either allele. For instance, those heterozygous for the gene for sickle-cell anemia have an enhanced resistance to malaria. Since males cannot be heterozygous for X-linked genes, only females can benefit from this phenomenon for genes on the X chromosome.

In past centuries the differential death rate of males and females was not so pronounced. Many women died in childbirth, or from infections due to childbirth, and from cancer of the breast or uterus. Now, however, since we have greatly reduced the toll from these events, the survival capacity of women has become manifest.

PROBLEMS AND QUESTIONS

1   From the discussion of abnormal sex chromosome distribution in this chapter, note that in addition to normal X-bearing eggs, females can produce XX eggs and no-X eggs. Males can produce normal X-bearing and normal Y-bearing sperm in addition to XY sperm, XX sperm, YY sperm, and sperm lacking both an X and a Y. Using a checkerboard, combine these various kinds of sperm and eggs. What combinations are possible? Can abnormal gametes ever give rise to normal fetuses? Would some of the combinations always be expected to abort? List the various combinations possessing a syndrome and name the syndrome.

2 For each chromosome combination included in the checkerboard you produced in answering Question 1, indicate the number of sex-chromatin bodies that one would expect to find in the individual's body cells.

3 On the basis of the fact that about 3 million babies are born in the United States each year, use the data on the sex ratio at birth (given in this chapter) to estimate how many males and how many females are to be expected in the total.

4 Some studies suggest that despite the larger number of live-born males, by the age of twenty the sex ratio has equalized. Speculate on the biological significance of this observation.

5 Assuming the autosomal chromosome number to be normal, indicate the *total chromosome number* and the *phenotypic sex* of each of the following individuals:
   (a) individual with Klinefelter syndrome who has two sex-chromatin bodies in each body cell
   (b) individual with Turner syndrome
   (c) individual with one sex-chromatin body and two fluorescent Y chromosomes per body cell
   (d) individual with no fluorescent Y chromosomes but with three sex-chromatin bodies per body cell

REFERENCES

BORGAONKAR, D. S., and S. A. SHAH. 1974. The XYY chromosome male—or syndrome? *Progress in Medical Genetics* 10:135–222.

MITTWOCH, U. 1963. Sex differences in cells. *Scientific American* 209(1): 54–62.

REYNOLDS, W. A., and T. R. MERTENS. 1964. The sex check. *American Biology Teacher* 26(6):411–15.

SHAPIRO, L. J., T. MOHANDAS, R. WEISS, and G. ROMEO. 1979. Non-inactivation of an X-chromosome locus in man. *Science* 204:1224–26.

WESTOFF, C. F., and R. R. RINDFUSS. 1974. Sex preselection in the United States: some implications. *Science* 184:633–36.

Heredity Influenced by Sex

THE DIFFERENCES IN THE CHROMOSOMES and hormones of the two sexes can influence the pattern of expression of many genes. In this chapter we shall consider such differences and patterns beginning with genes that lie on the X chromosome.

X-Linked Inheritance

Genes on the X chromosome have a pattern of inheritance and expression that differs from that of genes on the autosomes. For example, say that a man expresses a recessive gene on his X, but it is not expressed in any of his children. Then, some of the sons of his daughters show the trait, but it does not appear in any children of his sons. This has been called the **skip-generation method of inheritance.** Most of the genes on the X have no alleles on the Y. The lack of homology between the X and Y is apparent during synapsis of the chromosomes in meiosis. The X and Y do not pair throughout their lengths as the autosomes do, but only for a small region at one end of the pair (see Figure 6–1). This general lack of synapsis indicates that allelic genes are present only in that small region. A male, therefore, will express almost all the genes on his single X, including those that are recessive. Females, being diploid for all the genes on the X, will express only those recessive genes that are homozygous. The genes on the X that have no alleles on the Y are known as **X-linked genes,** but are often referred to by the older name, **sex-linked genes.** Let us illustrate with a well-known example.

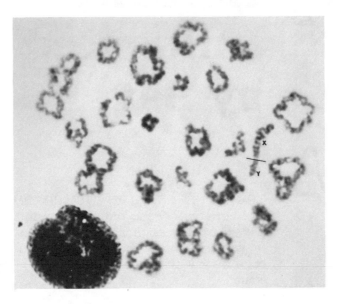

FIGURE 6—1. The end-to-end synapsis of the X and Y chromosomes during the first meiosis is shown clearly in this cell from a human testis. All the other chromosomes are synapsed side by side, thus indicating that allelic genes are present only at the ends of the X and Y. (Courtesy John Melnyk.)

Colorblindness

A boy went in a candy store and, pointing to a jar of suckers, asked for a green one. When the clerk took out a green one, the boy said, "No, not that red one. I want the green one." He then pointed to a red sucker. He thus discovered that he was colorblind, that he could not properly distinguish red from green. It has long been known that colorblindness is more common in males than in females, and, with the discovery of the sex chromosomes, the reason became apparent. The genes involved are on the X. The retina contains three basic kinds of cones that are sensitive to the three primary colors, red, green, and blue. The most common form of colorblindness is caused by a recessive X-linked gene that causes defective function primarily of the green-sensitive cones. The physiology of color vision is such that a weakness of green perception causes green to appear red.

Figure 6–2 shows how this type of colorblindness is inherited. A colorblind man will have the gene on his X and will pass it to all his daughters, but they will most likely receive the dominant allele for normal color vision from their mother and will have normal vision. Males do not receive an X from their father, however, so none of the sons will receive the gene. Half of the grandsons, through the daughters, however, will receive and express the gene. Females will be colorblind only if they receive the gene from both parents, so colorblind

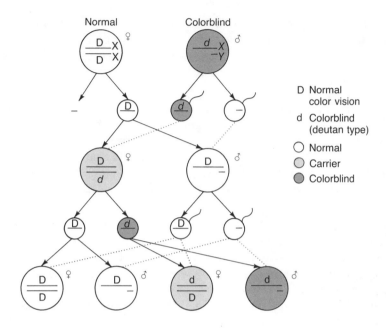

FIGURE 6–2. Pattern of inheritance of the X-linked gene for deutan (green-insensitive) colorblindness. Males express all the genes on their single X even though some of them may be recessive.

females are far less common than colorblind males. Since males can be neither homozygous nor heterozygous for X-linked genes, we use the term **hemizygous** to refer to their X-linked genes.

Those with defective perception of green have the **deutan** (sometimes called **deuteran**) **type** of colorblindness. Two varieties of the gene are known. The most common allele involves a mild defect in green perception, and red and green can be distinguished only if the light is good and the colors are bright. In poor light or with pastel shades, however, confusion is likely to arise. This is known as **deuteranomaly.** A less common allele causes an extreme defect in green perception in which red and green are likely to be confused under all conditions. This is called **deuteranopia.**

About three-fourths of those with red-green colorblindness have one of the deutan types; the others have the **protan type,** which involves cones that perceive red. This latter type also results in a confusion of red and green, this time because red objects appear green. One allele causes the mild **protanomaly,** while a less common allele causes the extreme **protanopia.** The genes for the protan type and the deutan type of colorblindness are both located on the X chromosome, but at separate loci. A much less common gene at a third locus on the X chromosome results in reduced blue sensitivity and the **tritan type** of colorblindness. This type is also found in two forms, **tritanomaly** and **tritanopia,** although the latter term is often applied to an autosomal dominant condition. There is also a very rare autosomal gene that causes total colorblindness, in which none of the cones is sensitive to color.

A girl may have normal color vision even though both parents are colorblind. The mother may be homozygous for the duetan type but have both normal alleles at the locus for the protan type. The father may have the protan type but carries a normal allele at the locus for the deutan type. Hence, all daughters would be heterozygous for both genes and have normal color vision (see Figure 6–3). The sons would all have the deutan type like their mother.

About 8 percent of the boys born in the United States have one of the two types of red-green colorblindness, but only about 0.4 percent of the girls are colorblind. This is about 1 boy in 12, but only 1 girl in 250. These percentages support the two-gene-locus concept. If only one locus were involved, we would expect 0.64 percent of the girls to be colorblind (0.08 × 0.08 = 0.0064). On the basis of two-gene loci, however, 6 percent of the boys would have the deutan type (three-fourths of 8 percent). About 0.36 percent of the girls would have this type (0.06 × 0.06 = 0.0036). Likewise, 2 percent of the boys would have the protan

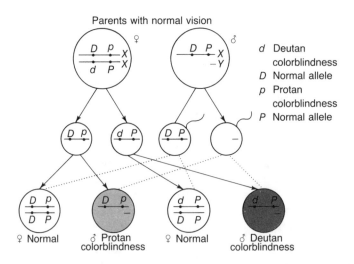

FIGURE 6–3. A woman who carries one gene for each of two kinds of colorblindness will have normal vision, but generally all of her sons will be colorblind. Some will have one kind and some will have the other. In rare cases she may produce a son with perfectly normal color vision or a son who has both kinds of color defective vision. These rare cases are produced by crossing over between the *D* and *P* gene loci on her X chromosomes.

type and 0.04 percent of the girls would have this type (0.02 × 0.02 = 0.0004). Adding the two percentages for girls, we get 0.4 percent, which is what is observed.

Hemophilia

Hemophilia, the bleeder's disease, is not common, afflicting only about 1 male in each 10,000, but its effects are so dramatic that it is well known. It results from a recessive X-linked gene that fails to produce a functional form of plasma protein known as **antihemophilic globulin (AHG),** or **substance VIII.** Without this globulin, normal blood clotting cannot take place. Hence, a person with this disease will bleed excessively even from minor injuries, and death often comes from excessive loss of blood. Much of the bleeding is internal, resulting from the rupture of small capillaries in the stomach or intestine, or it can be under the skin, causing large hematomas. This disease plagued the royal families of Europe. It seems to have appeared as a mutation in the immediate ancestry of Queen Victoria of England and through her descendants was spread to other royal families. The Tsarevitch Alexis of the Russian royal family had many close brushes with death because of hemophilia.

Those with hemophilia have a much greater chance of living normal lives today than in the past. The missing factor can be extracted from the plasma of blood donors and injected into those who lack it.

Hemophilia in girls is very rare. Since only 1 male in 10,000 has the gene, the chance that a woman will be a carrier is 1 in 5000 because she has two Xs. The random chance of pairing of a man with hemophilia and a woman carrier is the product of these two probabilities, which is only 1 in 50 million. Half of the daughters of such parents would get the gene from both parents, so the odds of expression would be 1 in each 100 million female births. However, this figure represents too high a frequency because many males with hemophilia die of bleeding before they are old enough to procreate children; at least they have in the past. Others may decide not to have children because of their affliction. Hemophilic girls have a serious problem at adolescence. They can easily bleed to death during a normal menstrual period. Some, however, have been known not only to survive adolescence, but also to bear children with the help of modern medicine.

Many factors are required for blood clotting, and excessive bleeding can result from defects in any of these factors. **Hemophilia A,** the classical hemophilia just described, is the most common affliction. Another gene at a different locus on the X chromosome causes a deficiency of another plasma substance, **plasma thromboplastic component (PTC).** This deficiency causes **hemophilia B,** or **Christmas disease,** which is not as serious as hemophilia A. There is also **pseudohemo-**

philia, which results from a dominant gene on an autosome. This gene interferes with the breakage of the blood platelets, or thrombocytes. Platelets contain a substance needed to initiate blood clotting. Still another gene, an autosomal recessive, interferes with the production of fibrinogen, a plasma protein that is converted into fibrin, which is the clot. Those homozygous for this gene have **afibrinogenemia.** These illustrations show how genes with different effects at the cellular level can result in the same phenotype on the organism level.

Until recently, classic hemophilia A could not be detected prenatally. Heterozygous women at risk for producing sons with hemophilia could have amniocentesis in order to permit the detection of the sex of their unborn child. If the fetus were female (and thus $X^H X^H$ or $X^H X^h$), the pregnancy would not be interrupted. If it were male, however, there was a 50 percent chance that he would have hemophilia ($X^h Y$). In cases where the fetus was found to be male, the couple could elect to terminate the pregnancy, knowing, of course, that half of the time the fetus would *not* have hemophilia ($X^H Y$).

An article in *The New England Journal of Medicine* in April, 1979, announced the successful prenatal diagnosis of hemophilia A. Fetal blood plasma obtained by fetoscopy can be studied using radioimmunoassay techniques to determine the presence or absence of factor VIII. Only those male fetuses having hemophilia would then be considered for abortion.

X-Linked Traits Expressed Only in Males

It is a common error to think that X-linked traits are expressed in only one sex. Although we have seen that this is not true for many of these genes, there are a few X-linked recessive genes that are never expressed in females because males who carry these genes never live to the age of sexual maturity and therefore never pass them on. One such gene causes **pseudohypertrophic muscular dystrophy,** also called the Duchenne type of muscular dystrophy. A boy with this gene has a normal early childhood, but at about six or seven years his muscles begin to swell. This is followed by a gradual deterioration and shrinkage until by the early teens he is almost literally skin and bones (Figure 6–4). Death follows shortly. A less common form of muscular dystrophy is caused by an autosomal recessive gene and can afflict both sexes equally. X-linked, recessive, lethal genes that cause intrauterine or neonatal death also would be in this category of genes expressed only in males. Half of the sons of a woman carrier of such a lethal would not survive, but none of her daughters would be affected.

In an effort to assay the possible harmful genetic consequences of exposures to high-energy radiation, some geneticists have turned to the variations in the sex ratio among children of exposed parents. As

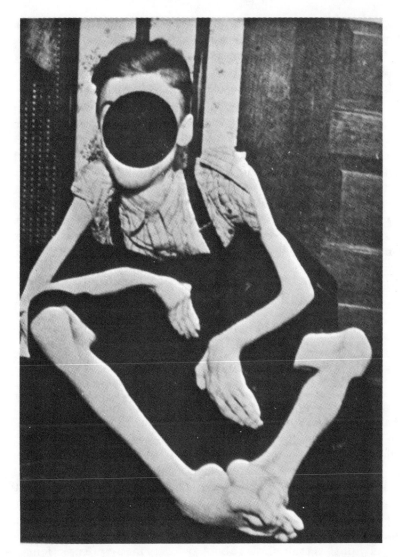

FIGURE 6–4. A boy with advanced pseudohypertrophic muscular dystrophy, a result of an X-linked recessive gene received from his mother.

has already been pointed out, lethal mutations are more common than those that have a visible effect. A recessive lethal induced on an X chromosome will cause the death of any male child, but females will not be harmed. Hence, any agent that increases the mutation rate will tend to reduce the number of sons born to women exposed to the agent. A small proportion of the lethal mutations are dominant, and if such a mutation is induced on the X chromosome of a man, all daughters receiving it will die. Dominant lethals induced on the mother's X chromosome will kill both male and female fetuses, in equal numbers, so will not affect the sex ratio. The X is a medium-sized chromosome out of a total of twenty-three kinds in each cell, so the total number of lethal mutations induced on all chromosomes must be much greater.

To determine the possible mutagenic effect of X rays on humans, Raymond Turpin in Paris studied over four thousand patients who had received extensive X irradiation for gastrointestinal examinations and other extended examinations and treatments. The dosage received was from about 70 to 270 rads, which is over a hundred times the amount a person receives from a simple chest X ray. The results are shown in Table 6–1. The decrease in the percentage of male births after radiation of women and the decrease in percentage of female births after radiation of men indicate the possible induction of lethal mutations on the X chromosome.

TABLE 6–1   Effect of X rays (70–270 rads) on offspring

	Percentage of Males in Offspring Before Radiation	Percentage of Males in Offspring After Radiation
women	54.1	48.5
men	51.5	56.1

X-Linked Traits More Common in Females

Some genes on the X chromosome are expressed more frequently in females. These are dominant genes. Females, with their two Xs, have twice the chance of expressing such genes. **Defective dentine** of the teeth is caused by a dominant gene on the X. The gene causes the teeth to wear rapidly, leaving only stumps by adolescence. A man with this trait will pass the gene to all his daughters, but to none of his sons. A woman with it will pass it equally to both sexes. A tabulation of the frequency of this trait shows about twice as many affected females as males.

Heterozygous Expression of X-Linked Genes

We learned in Chapter 5 that only one X uncoils and functions in a female. The other remains tightly coiled in the Barr body. Since different Xs are inactivated in different cells, however, some cells of a woman express the genes on one X and other cells express the genes on the other X. This can sometimes be demonstrated by an analysis of individual cells. **Favism** is a common affliction in Sicily, in southern Italy, and in other Mediterranean areas of Europe. Individuals with this condition will become anemic if they eat fava beans (broad beans similar to lima beans). They will even have anemia after working in fields where the bean plants are flowering because they inhale the pollen. Also, certain drugs cause the anemia. These include primaquine, an antimalarial drug, sulfanilamide, and mothballs, which are sometimes eaten by children. The red blood cells have a deficiency of an enzyme, **glucose-6-phosphate dehydrogenase,** usually abbreviated to **G6PD.** Red blood cells that have the X-linked gene for this condition cannot produce the enzyme and this interferes with a minor biochemical pathway in glycolysis. As a result, a particular compound needed in the red blood cell membrane is missing, and the cells become very fragile, many break (become hemolyzed), and the anemia develops. Those persons with the gene may have an advantage, however. They appear to be more resistant to malaria, which was once very prevalent in the regions where favism is common. Selection favored those with the gene as long as they avoided the beans.

Heterozygous women also may develop a mild anemia when they eat the beans because some of their cells lack the enzyme. Nile blue sulfate is a stain that will turn cells blue if G6PD is present. Some of the heterozygous female cells will turn blue and some will not. Some of the stem cells in the bone marrow that produce red blood cells will include an active X with the gene for enzyme production and some will not. The proportion of cells that will turn blue varies because of the chance inactivation of the X in early embryonic development, but it will average about one-half. Hence, heterozygous women will have an anemia when exposed to the offending substances, but to a milder degree than men who are hemizygous for the defective allele.

Some women who are heterozygous for the hemophilia allele have a delayed clotting time of the blood. Their plasma may have as little as 20 percent of the antihemophilic globulin present in women who are homozygous for the normal gene. Other heterozygous women may have 80 percent or more of the normal quantity of this factor. In the first group the active X in most of the cells that produce AHG will carry the gene for hemophilia, while in the second group the active X in most cells will carry the normal allele. For the same reason, a few of the women who are heterozygous for colorblindness may have some difficulty distinguishing red and green in one or both eyes.

Children with **Lesch-Nyhan disease** have defective purine metabolism and produce an excess of uric acid. This causes brain damage, spastic muscle movements, and a desire for self-mutilation. They may bite their lips and hands and have to be restrained to prevent serious injury to themselves and those around them. A recessive X-linked gene is responsible, and heterozygous females can be recognized by a study of the roots of hairs plucked from the body. Since a hair root is often only one cell type and because of the random inactivation of the Xs, about one-half of the hair roots of heterozygous females will have the enzyme for normal purine metabolism while the rest will not. This test is often made on women from families in which the gene has been expressed to determine if they are carriers.

According to the Lyon hypothesis, the inactivation of an X chromosome occurs early in embryonic life, so the islands of tissue that express an X-linked recessive must be rather large. Evidence for this is found in the expression of the gene for **anhidrotic ectodermal dysplasia.** Men hemizygous for this gene do not have any teeth and lack sweat glands in the skin. Heterozygous women may have areas of the jaws without teeth and large islands of the skin without sweat glands. The extent and location of the areas varies with different women, as would be expected.

Genes on the Y Chromosome

Most of the Y chromosome does not contain genes. It is heterochromatic, as indicated by the large area of fluorescence under ultraviolet light. Still, we know that it contains a gene that activates the testicular portion of the primitive gonad. Also, since the X and Y synapse during meiosis, the Y chromosome must have some genes that are alleles of some of those on the X because only allelic genes are attracted to each other at this time. The end-to-end pairing of the X and Y chromosomes suggests that there must be only a few genes at the end of the long arm of the X and the short arm of the Y that are alleles.

Y-Linked Inheritance

Since most of the Y has no alleles on the X, males will be haploid for all the genes that lie on this nonhomologous portion. Such Y-linked or **holandric** genes have a unique pattern of transmission. Men will express any genes in this region, and the genes will be passed from father to son, but never from father to daughter. Many potential Y-linked traits have been proposed, but most of these have been shown to be due to chance assortment of genes at other locations. Most authorities seem to agree that a gene for a specific male antigen, H-Y, is holandric, as is a testis determining factor (TDF). Some investigators suggest that these two phenotypic effects may actually be controlled by one holandric gene. Another gene that has some evidence to support Y linkage is **hairy pinna,** long hair growing on the ears. In certain families in India this trait has appeared in all males who have fathers with it, but is never transmitted or expressed by women.

Homologous Genes on the X and Y

In some experimental animals, such as fruit flies, it has been possible to demonstrate inheritance of genes on the homologous regions of the X and Y. These genes, called incompletely sex-linked genes, are diploid in both males and females and would be transmitted in a manner similar to autosomal genes, but are identified by linkage studies that associate them with genes on the X. Such inheritance has been suggested for several human traits, but no conclusive evidence has substantiated the suggestions.

Sex-Influenced Traits

A number of traits in some of the nonhuman forms of life have been found to be dominant in one sex and recessive in the other. The genes involved are characteristically autosomal, but their expression is affected by the internal hormonal environment of the individual in which they occur. The inheritance of **pattern baldness** is usually described as being sex influenced, although some investigators believe it to be regulated by a sex-limited gene as described in the next section of this chapter. Individuals exhibiting pattern baldness suffer the loss of hair, first on the front and top of the head but not on the sides, beginning during the middle years of life. The transmission of pattern baldness in the descendants of U.S. President John Adams supports the view that the condition is regulated by a single gene.

If we use B' to symbolize the allele for pattern baldness and B for the allele for a normal head of hair, then, if the condition is sex influenced, the following genotypes and phenotypes will occur:

	male	female
BB	normal	normal
BB'	bald	normal
$B'B'$	bald	bald

Note that a heterozygous man becomes bald, but a heterozygous woman does not. This is consistent with sex-influenced inheritance with B' being dominant in the male, but recessive in the female.

Another human trait that is said to be sex influenced in inheritance is the length of the index finger. If the tip of the ring (fourth) finger is placed on a line, the index finger may or may not reach the line (see Figure 6–5). Short index finger is thought to be dominant in males and recessive in females as follows:

	male	female
SS	long	long
SS'	short	long
$S'S'$	short	short

Note that the gene is thought to be autosomal with the S' allele for short index finger expressing in heterozygous males but not in heterozygous females. In a family where the husband has long index finger (SS) and the wife has short index finger ($S'S'$) all children will be

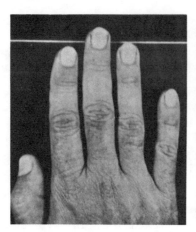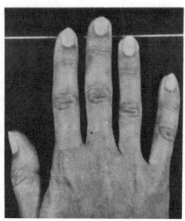

FIGURE 6—5. The length of the index finger in relation to the fourth finger appears to be sex-influenced. An index finger shorter than the fourth finger seems to be dominant in males, but recessive in females. The male hand on the left has the short index finger, and the female hand on the right has the long index finger.

heterozygous. Hence, all sons will have the short finger and all daughters the long finger.

Sex-Limited Genes

Sex-limited genes are those that are normally expressed in only one sex, although they are carried equally by both sexes. They may be on any chromosome, but are mostly on autosomes since the X carries only a small proportion of the total genes in a cell.

Most sex-limited genes code the expression of traits related to sex characteristics. These include the primary sex traits, which are related directly to reproduction, and also secondary sex characteristics, such as growth of hair on the face and differences in skeletal and muscular development. A woman does not normally have a beard, yet she carries all the genes necessary to produce a beard. The beard types of her sons depend on genes they receive from her just as much as those from the father. On rare occasions, some of these genes may be expressed in the wrong sex. The bearded lady, often on exhibition in carnival shows, may be a normal woman in other respects, but her genes for extensive hair growth on the face have escaped inhibition. Also, in otherwise normal men, breast enlargement, **gynecomastia,** may occur.

As explained above, pattern baldness is thought by some workers to be an example of a sex-influenced trait. Others have claimed, however, that it is sex limited, with baldness developing in males only, as follows:

	male	female
BB	normal	normal
BB'	bald	normal
B'B'	bald	normal

Note that according to this hypothesis, the allele for baldness (*B'*) acts as a dominant in the male, but is never expressed in the female, despite the fact that she may possess and transmit it.

A man with low testosterone production may be somewhat feminized because, like all men, he produces some estrogen from his adrenal glands. He may take testosterone to increase his masculinity, but will begin to lose his hair. Women with some types of cancer may be given testosterone to reduce their output of estrogen because estrogen stimulates the growth of some types of cancer. They may begin to lose their hair if they carry the genes for baldness. Studies by Ivor Mills in England showed that many women working in occupations involving a great deal of mental stress may also begin losing their hair as a result of hormone changes that accompany the stress. Hair growth may increase on the face and chest as well. Similar effects have been observed among women subjected to the stress of confinement in mental and penal institutions.

PROBLEMS AND QUESTIONS

1 A woman with normal color vision married a man with deutan colorblindness. They had one son who also had deutan colorblindness. Is it correct to say that the son inherited his father's vision defect? Explain.

2 A woman with deutan colorblindness married a man with protan colorblindness. They had two children—a daughter with *normal* color vision and a son with deutan colorblindness. Explain the inheritance of color vision in this family by writing the genotypes of all individuals involved.

3 A man was found to possess an unusual trait that was seemingly inherited. His wife did not have this trait. The man transmitted the trait to all of his daughters but to none of his sons. His affected daughters married unaffected men but transmitted the trait to approximately one-half of their children of either sex. What appears to be the mechanism of inheritance of this trait? Explain by assigning genotypes to all individuals involved.

4 A woman unaffected with pattern baldness married a man who became bald in his late twenties. They had two children—a daugh-

ter who was affected with pattern baldness and a son who was not. Explain the mechanism of inheritance of pattern baldness in this family and assign genotypes to all individuals involved.

5 Complete the following matching exercise.

_____ (1) A trait that is always transmitted from father to son but never occurs in females has what mechanism of inheritance?	**(a)** sex-limited **(b)** sex-influenced **(c)** X-linked **(d)** Y-linked **(e)** hemophilia **(f)** pattern baldness **(g)** H-Y antigen **(h)** beard
_____ (2) An autosomal trait that is expressed in one sex only but can be transmitted by either sex is said to be _____.	
_____ (3) A gene on the X chromosome does not have an allele on the Y chromosome; a trait that is due to the gene is said to be _____.	
_____ (4) An autosomal gene can be transmitted by and expressed in both sexes, but one allele acts as a dominant in the male and the alternate allele acts as a dominant in the female; this gene is said to be _____.	

REFERENCES

FIRSHEIN, S. I., et al. 1979. Prenatal diagnosis of classic hemophilia. *The New England Journal of Medicine* 300(17):937–41.

HAMMOND, R. E. 1980. Human color vision. *Carolina Tips* 43(12):55–57.

McKUSICK, V. A. 1965. The royal hemophilia. *Scientific American* 213(2):88–95.

PHILPS, V. R. 1952. Relative index finger length as a sex-influenced trait in man. *American Journal of Human Genetics* 4:72–89.

RUSHTON, W. A. H. 1975. Visual pigments and colorblindness. *Scientific American* 232(3):64–74.

SNYDER, L. H., and H. C. YINGLING. 1935. The application of the gene-frequency method of analysis to sex-influenced factors, with special reference to baldness. *Human Biology* 7:608–15.

Immuno-genetics

WHY ARE SOME PEOPLE EASILY INFECTED with disease organisms while others, equally exposed, resist infection? Why do some people have hay fever when certain pollens are in the air while others, inhaling the same air, have no such difficulty? Why can you accept blood transfusions from some persons, but not from others? What is the role of heredity in all these reactions? Answers to these important questions are to be found in an understanding of the immune system, a system strongly influenced by heredity. The study of this relationship is known as **immunogenetics.** Antigens and antibodies play a major role in the immune system and we must understand these before we can proceed with a study of the role of heredity.

Antigens and Antibodies

Antigens are a basic part of the molecules of all proteins, some carbohydrates and lipids, and even some inorganic molecules, such as mercury compounds. They are characterized by their ability to stimulate antibody production when they enter the body of higher vertebrate animals, especially the birds and mammals. **Antibodies** are proteins in the gamma globulin fraction of the blood plasma and are also known as immunoglobulins (Figure 7−1). They have the ability to react with and alter substances containing the antigens that stimulated their production. This can be better understood by an example in an experimental animal.

If a guinea pig is injected with egg albumen, no reaction will occur if this is the first such injection. However, the guinea pig's immune system will react to this foreign protein by producing antibodies against the antigens in the protein. These antibodies will combine with the albumen, and it will be eliminated. The antibodies continue to be produced and will be present in the blood plasma. If, after two or more weeks, another large

105

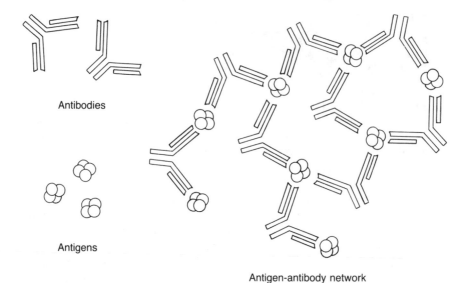

Antibodies

Antigens

Antigen-antibody network

FIGURE 7–1. Antibodies (immunoglobulins) bind to specific antigens, forming a network of antigen-antibody components, and thereby inactivating the invading antigen.

injection of albumen is given, a severe reaction occurs as the antibodies combine with the antigens of the albumen. The guinea pig may go into spasms and even die as a result. Anyone who has had a reaction to a penicillin injection can testify to the severity of a strong antigen-antibody reaction.

Antibodies serve as a protection against infections. When bacteria or viruses enter your body, you make antibodies against the antigens that are a part of these invaders. The antibodies not only help you to overcome the initial invasion of organisms, but remain in your blood to destroy the microorganisms and prevent future infections. You develop an immunity to this type of infection. Through vaccination, the introduction of weakened or killed microorganisms, we can stimulate antibody production and give protection without an initial infection. Some people, however, cannot produce these vital antibodies. They have inherited a recessive condition known as **agammaglobulinemia;** they cannot produce gamma globulin, the part of the blood that includes the antibodies. Unless they are kept isolated in a germ-free environment, they usually die early as a result of what would otherwise be minor infections.

Antibodies are highly specific: those produced in response to one type of antigen are ineffective against other antigens that may be only slightly different. A mutation that alters the protein structure of a disease organism may make it highly infective, because it is not attacked by antibodies produced during previous infections or vaccinations. The frequent appearance of a new strain of the influenza virus is a good example of such mutation. The mutation usually occurs in a single virus particle in a certain region, and then spreads over the world quickly because of the high mobility of our present populations.

Not everything we can say about our immune system is good. The same type of reaction that protects us from invading microorganisms can sometimes cause us quite a bit of misery. Hay fever, allergic asthma, hives, and other allergic reactions are the result of antibody production against antigens in our environment. Also, while the surgery necessary for successful organ transplants can be done with relative ease, the prevention of rejection of the transplants is very difficult. Our immune system recognizes the antigens of the transplant as foreign and antibody production will cause a rejection unless **immunosuppressants** (substances that suppress antibody production) are given. Identical twins are the only ones that have identical protein structure in their body organs and, hence, are the only ones that can exchange transplants without such suppression.

The Immunological Reaction

The immune system is quite complex and not yet completely understood, but recently developed techniques have given some insight into the events involved. Some of the leukocytes (white blood cells) play an important part.

Leukocytes Involved in the Immune Reaction

Three kinds of leukocytes are involved in the immune reaction (Figure 7–2). Two of these are **lymphocytes,** which have a single, large nucleus, in contrast to the **granulocytes,** which are larger, have granules in the cytoplasm, and have smaller nuclei with two or three lobes. There are two kinds of lymphocytes, known as **B** and **T lymphocytes,** that appear the same under the microscope, but can be distinguished by special tests. Precursors of both of these are produced in the bone marrow of the fetus and are released into the fetal circulatory system. Those destined to form B lymphocytes settle in the cortex of the lymph nodes and the spleen. There they undergo continuous reproduction throughout the rest of life and release mature B lymphocytes into the vascular system. Cells that are to form the T lymphocytes, on the other hand, migrate to the thymus gland at the front of the chest. They apparently are retained in the thymus until after birth and then migrate to the lymph nodes and spleen. There they reproduce and are released as mature T lymphocytes.

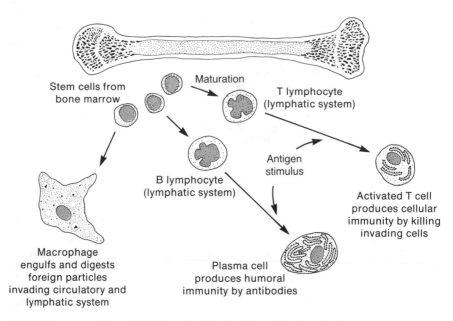

Stem cells from bone marrow

Maturation

T lymphocyte (lymphatic system)

Antigen stimulus

B lymphocyte (lymphatic system)

Activated T cell produces cellular immunity by killing invading cells

Macrophage engulfs and digests foreign particles invading circulatory and lymphatic system

Plasma cell produces humoral immunity by antibodies

FIGURE 7–2. The white blood cells that function in the immune response trace their origin to the bone marrow.

Macrophages, the third type of leukocyte involved, are very large leukocytes produced in the bone marrow, where they remain until released in a mature form. They are highly amoeboid and can squeeze through very small places, even leaving the blood stream and moving about in the intercellular spaces. They can recognize and engulf foreign particles. Their cytoplasm contains many **lysosomes,** which are small packets of enzymes, that can release their contents when they are needed to digest the engulfed particles.

Role of Leukocytes in the Immune Reaction

The collaboration of all three of these kinds of leukocytes is necessary for an effective immune reaction. The process, as we now understand it, is something like this. Assume that some pathogenic bacteria enter your body. Some of them will soon contact and adhere to the surface of some of the macrophages. While here they will also be contacted by B and T lymphocytes, which recognize them as foreign. The combined contacts stimulate the leukocytes to respond. Some of the B lymphocytes change into plasma cells that begin producing antibodies that are specific antagonists of the antigens of the bacteria. The plasma cells multiply and continue producing the antibodies, which are released into the blood stream. These circulating antibodies can react with and help destroy the bacteria of this kind, a condition known as **humoral immunity.**

Antibodies in the plasma can react with and neutralize foreign substances in several ways. Some are **agglutinins;** they cause the foreign cells to aggregate in large clumps where they can easily be engulfed by the macrophages. Each antibody molecule has at least two receptor sites and when these become attached to two different cells, the cells are bound together. Antibodies of the different ABO blood types function this way and cause the red blood cells to become agglutinated when incompatible bloods are mixed (see Chapter 8). When the protein molecules are dispersed in a liquid, the antibodies may cause them to be bound together in large groups and the mixture becomes cloudy. These antibodies are **precipitins.** Other antibodies are **lysins;** they destroy the cell membranes of foreign cells and cause the cell contents to be lost.

T lymphocytes become programmed to be **killer cells.** They produce and retain antibodies on their surface and can attack and destroy bacteria with this particular kind of antigen by contact. This reaction is known as **cellular immunity.** These killer cells accumulate in large numbers at the site of an infection and help destroy invading organisms. They are abundant in pus and give it its whitish appearance. This type of reaction accounts in a large measure for the rejection of organ transplants. Immunosuppressants inhibit the development of killer cells and prevent the rejection of the implant, but will also reduce the resistance to infections.

The macrophages complete the triple threat to invaders. These large cells are stimulated to become particularly aggressive against cells with antigens that stimulated the reaction. They multiply rapidly upon infection and engulf the invaders in large numbers.

Continuing Immunity

Once an infection has been overcome, the plasma cells and killer cells remain in the blood stream for a time and can destroy any additional microorganisms that may enter the body if they are of the kind that stimulated the reaction. The number of these cells will gradually decline, however, until additional infection with these organisms is again possible. This time, however, the body is ready; some of the lymphocytes have not become plasma cells or killer cells after stimulation, but remain as "memory cells." They retain the information about the foreign antigens and are quickly converted into plasma cells or killer cells when again confronted by the same kind of invaders. This suggests why booster shots are effective after a vaccination. The vaccination stimulates the immune system, but after that only a very small amount of the antigen is needed to reactivate the memory cells and restore full immunity.

The length of time of complete immunity after an infection or vaccination varies considerably with different diseases. For some it is many

years, but for others it barely lasts for a season. Generally this is correlated with the incubation period of the disease, the time between the
first exposure and the development of symptoms. If a disease has a
long incubation period, the memory cells have time to become activated
and respond before symptoms appear, so the immunity is long lasting.
In diseases with a short incubation period the symptoms appear before
the memory cells can be fully activated, and the immunity may be of
short duration.

Age and Antibody Production

A fetus does not produce antibodies for reasons we shall explain later,
but a newborn baby will have antibodies in the blood stream. These
have come from the blood of the mother and have passed across the
placenta and give the newborn some protection against the many
harmful microorganisms that will be in the external environment.
Like all antibodies that are placed in a body from an outside source,
however, these disappear within a few weeks to a few months after
birth. At about two months of age, however, a baby's own immune
system will begin operating and will reach its maximum efficiency at
about eight to ten years of age. It retains a high efficiency through
youth but begins to decline at middle age, and in old age the antibodies
produced in response to infection may be only one-fourth those produced by a young person. This can account for the increasing severity
of infections and high death rate from infections in old age. It can also
explain the increased incidence of cancer in older people, because with
aging our antibodies no longer give a high degree of protection from
the development of malignancies.

Inheritance and the Immune Reaction

The Physical Nature of Antibodies

Antibodies, or immunoglobulins, vary in molecular weight and arrangement of amino acids. Through electrophoresis they can be separated into a number of different types. These are designated as IgA,
IgD, IgE, IgG, and IgM. IgA may be present in body secretions, such as
mucus, saliva, and intestinal secretions. Those who suffer from allergies can blame their IgE. IgM is involved in the production of antibodies against large invading cells. IgG, however, is the most common and
important in most immunities and has been the most widely studied. It
has a molecular weight of 165,000, which is small enough to allow it to
pass through the placenta and transfer a mother's immunity to her
fetus.

Each molecule of IgG is made of two heavy and two light polypeptide
chains of amino acids (Figure 7–3). Each chain consists of a constant

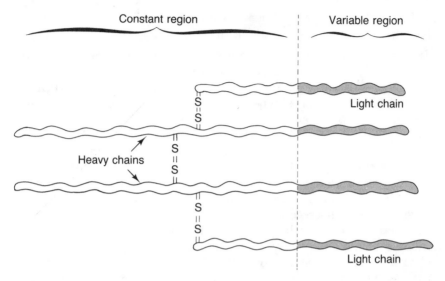

Constant region | Variable region

Light chain

Heavy chains

Light chain

FIGURE 7–3. The immunoglobulin molecule is made of four chains of amino acids, two heavy and two light. Both types of chains have a constant region and a variable region.

and a variable region. The constant region of the heavy chains is coded by the **Gm locus.** Over twenty-five different alleles have been found at this locus. The constant region of the light chains is coded by the **Inv locus** and three alleles have been found at this locus. The variations in these chains influence a person's ability to react to antigens. Some alleles may result in molecules that react readily and produce abundant antibodies, while those with another allele may respond more slowly and produce a lower titer of antibodies. This can account for inherited differences in susceptibility to diseases and the ability to recover from infections readily.

Theories to Explain the Variety of Antibodies

The number of different kinds of antibodies in the blood plasma of a person must be enormous, considering the many different foreign antigens that we contact. How can we produce so many antibodies from the same basic inherited immunoglobulins? One possibility is that the variable chains of the molecule are all the same when the molecules are produced, but are altered in their amino acid sequence when they contact specific antigens. Such an alteration would make the immunoglobulin molecule an antibody against this antigen. Furthermore, this molecule could serve as a template for the production of more antibodies of the same kind. This would be a unique situation, one in which the amino acid sequence of a gene product could be changed by environmental forces and the gene product reproduce itself in the altered form. Since this seems highly unlikely, other possible explanations have been sought.

In recent years a somewhat more plausible **clonal selection theory** has been proposed. This theory holds that the B lymphocytes are produced in great variety, one type for each of the many different kinds of antigens to which a person can react. Studies on mice have located a region of a chromosome that produces the variable parts of the IgG molecule. This region is large enough to contain many different genes, enough to account for the great variety of different amino acid combinations on the variable chains. A different codominant gene codes each of the different combinations. Some mice were found that could not produce antibodies against certain antigens no matter how great the exposure. These mice evidently did not have the genes for the production of variable chains that could react with these antigens and form antibodies. This must be the case when certain people cannot produce antibodies against certain antigens. Most people in certain populations, for instance, seem to lack the ability to produce antibodies against the antigens of certain disease organisms. As a result, they may suffer greatly and many of them may die from infections that are rather minor in other populations. Native Hawaiians died in great numbers when the missionaries brought the measles virus to their islands. Never having been exposed to this virus before, there had been no selection for genes that would produce B lymphocytes programmed to produce antibodies against the virus.

All people lack the ability to produce antibodies against certain antigens. People have antigens known as M, N, or both on the surface of their red blood cells, but those who lack one of these antigens cannot produce antibodies against it. Guinea pigs, however, can produce such antibodies, so they must have a gene for a kind of B lymphocyte that we lack. For other blood antigens, only a few people can produce antibodies. See Chapter 8 for details about these.

Distinction Between Self and Nonself

Each person has many antigens as a part of the body, and if antibodies were produced against these the reaction would destroy the body tissues. This obviously does not happen, yet how can the immune system distinguish between antigens that are a part of one's own self and those that are foreign or nonself?

Fetal Reactions

The clonal selection theory offers a possible explanation for the body's ability to distinguish self from nonself. Antibodies are not produced by a fetus. This is necessary because if antibodies were produced, the fetus might produce them against the cells of the mother's uterus, which are so closely intertwined with the placenta, and cause a rejection of this life-support system. This inability to produce antibodies seems to be because the T lymphocytes are not released from the thymus gland before birth, at least not in functional form. The mature B lymphocytes are released and macrophages are present, but without T lymphocytes to complete the three-part immune reaction, no antibodies are formed. The B lymphocytes, however, contact antigens of the body and those that would produce antibodies against them are inactivated. Hence, they are not present in active form to function when the immune system becomes fully functional after birth.

Support for this concept is provided by experiments on mice. If newborn mice are given injections of cells from other mice, they will later accept transplants from these donors as if the tissues were their own. Apparently, the B lymphocytes that would react with these foreign antigens are inactivated and are not able to react to them later. Human observations give further support for the concept. It has long been known that identical twins can give transplants without immunosuppressants because such twins have the same antigens. Recently, however, it has also been found that dizygotic twins can sometimes give transplants to one another without any reaction. Dizygotic twins have differences in some of their genes and this means that they will have differences in some of their antigens, so how can this be explained? These twins share the same uterus before birth and sometimes they may develop a connection with one another. Some blood and other tissue may cross this connection and the fetal immune system learns to recognize this tissue as self. As a clinical application of this principle, it has been suggested that we might give newborns an injection of mixed human tissues so that they will have a tolerance of many human antigens that are not a part of their own body. This would make possible a much greater chance of success of transplants if they should ever be needed in later life.

Autoimmunity

Sometimes the delicately balanced immune system makes a mistake and antibodies are developed against the body's own antigens. The results can be various debilitating conditions, the nature of which depends on the particular kind of tissues with which the antibodies react. This is known as **autoimmunity** and has been found to be the cause of many previously unexplained human defects.

Rheumatic heart disease is an autoimmune disease that is fairly common. It appears most often in children after a severe throat infection by a certain strain of streptococcus bacteria. By chance an antigen on the surface of these bacteria is very similar to an antigen on the myocardium, the muscle tissue of the heart. It is known that when antibodies reach a very high titer in the blood they can react to some extent with antigens that are similar to the antigens that stimulated their production. Hence, if a child has a very efficient immune system and produces a high titer of antibodies against the streptococcus, and also has inherited individual antigens of the myocardium that are very similar to those of the bacteria, the antibodies may react with the myocardium and cause the heart difficulties. In other persons it is the cells of the kidneys that have an antigenic similarity to those of the bacteria and **nephritis** may follow a streptococcus throat infection.

Many other diseases that have at least some relation to autoimmunity have been identified; we now realize that this misplaced immune reaction is a major cause of many human ailments. The partial list given below will give you some idea of the extent of the problem:

Ankylosing spondylitis is characterized by a gradual fusion of the vertebrae to give a stiff spine, often known as poker spine. Antibodies are formed that react with the cartilages between the vertebrae and cause irritation and pain when the back is moved. Calcium is deposited in these irritated cartilages that then fuse with the bone to give a spinal column of solid bone.

Hodgkin's disease is an enlargement and inflammation of the lymph nodes as a result of antigen-antibody reaction at these nodes. Formerly nearly always fatal, treatment that can prolong life is now available.

Myasthenia gravis is an antibody reaction against antigens of the neuromuscular junction, with a resulting inactivation of muscles.

Hashimoto thyroiditis is an antibody reaction with the thyroid gland.

Reye syndrome is an inflammation of the brain that sometimes follows an influenza infection, particularly in children. It is often fatal.

Rheumatoid arthritis is a painful inflammation of the joints that afflicts many people especially in middle and old age.

Systemic lupus erythematosus, often shortened to lupus or SLE, is one of the most common of the autoimmune diseases, yet is often not recognized because of its many manifestations. It results from antibodies produced in response to the nucleic acid of viruses that can cause cancer. These nucleic acids vary greatly and may be similar to the antigens of different parts of the body. (See Chapter 14 for a discussion of the role of viruses in cancer.) The reactions may be with the skin, joints, kidneys, lungs, heart, and even the brain.

Multiple sclerosis is associated with sclerotic (calcified) patches that appear around nerves in the brain and spinal cord. It causes a gradual and prolonged loss of muscle control with eventual death.

Male sterility can result when a man develops antibodies against the antigens of his sperm and renders them incapable of fertilization.

Addison's disease involves the destruction of some of the tissue of the cortex of the adrenal glands, causing a deficiency of their hormones, such as cortisone.

Pernicious anemia is the antibody destruction of cells of the stomach that produce the intrinsic factor needed to stimulate the release of red blood cells from the bone marrow.

Hemolytic anemia involves the antibody destruction of the membranes around red blood cells, causing them to break and release the hemoglobin.

Ulcerative colitis is the antibody irritation of the lining of the colon.

Paralytic polio is caused by an antibody reaction with the nerve connections. Many people once had polio, but only a few had the paralysis that followed it.

Juvenile diabetes often follows an attack of mumps or scarlet fever. The islands of Langerhans that produce insulin in the pancreas are injured so that sugar diabetes results. Insulin injections are needed to permit proper utilization of dietary carbohydrates.

This incomplete list shows how extensive autoimmunity can be. It is a major medical problem today. Some progress has been made in overcoming some of these diseases, but they are very difficult to treat because of the nature of the autoimmune reaction. As we shall learn in the next section, heredity plays a major role in determining our overall susceptibility to autoimmune diseases and the particular kind that we might develop.

Sex plays a role in autoimmune diseases. Women are much more likely to be the victims of most of these diseases. For instance, about 80 percent of the victims of SLE are women, and about seven out of each eight cases of Hashimoto thyroiditis are women. Male sex hormones seem to give males some protection against autoimmune diseases because castrated male animals have about the same incidence as females of the same species. Poker spine seems to be an exception with more men than women having this disabling condition, although tests showing minor cases place the frequency at about equality.

Human Leukocyte Antigens

When it was discovered that the leukocytes had a great variety of antigens on their surfaces, it was also found that these were correlated with susceptibility to specific autoimmune diseases. These antigens, commonly known as HLA, show considerable variation as a result of different alleles at several loci on the chromosomes. They first came to the attention of physicians in connection with transplants. When donors and recipients had closely matching HLA types, the transplant was much more likely to be successful than when the HLA types were quite divergent. Hence, the HLA became known as the histocompatibility antigens. They were found on other tissues besides the leukocytes, and that is why they give a good indication of compatability of transplant donors and recipients.

Discovery of HLA and Application to Transplantation

The HLA antigens were discovered when it was found that some of the rather mild reactions following blood transfusion were the result of a reaction of the serum of the donor with the leukocytes of the recipient. This was because the recipient had received a transfusion of a foreign HLA type previously and developed antibodies against the antigens on the leukocytes. Also, when a woman carries a fetus with a foreign HLA type (inherited from the father) some of the leukocytes from the fetus can enter her body and sensitize her. She could then have some reaction to transfusion of blood from the HLA type that sensitized her. Or some of her antibodies might pass through the placenta and cause difficulties in the fetus of a future pregnancy. These reactions are usually mild and not of great clinical significance, but they have given us information on the HLA types. Blood sera from those who have become sensitized, either by transfusion or pregnancy, can be used to identify the HLA types.

The HLA antigens are determined by four loci, designated as A, B, C, and D, that are very closely linked on chromosome 6. The variety of HLA types is very great because there are multiple alleles at each of the loci. For instance, thirteen alleles have been identified at the A locus and twenty-two at the B locus. As a result, a complete matching of the HLA types for transplants is practically impossible, except in the case of identical twins. However, some of the HLA types are more common than others (see Table 7–1), and many of these can be matched. A computer network linking large hospitals all over the country stores data on the HLA types of those individuals needing a transplant and of the organs that become available. Let us say that a kidney becomes available in Seattle and the computer shows that, of all the people needing a kidney transplant, a woman in Boston matches the HLA types most closely. The kidney is sent to Boston by air and the transplant is made. Close relatives will match more closely than nonrelated persons because of the common genetic ancestry, and those of the same racial type will match more closely than those of different races. The proportion of HLA types varies considerably in different races.

TABLE 7–1 Frequency of HLA-A and HLA-B alleles (From study by Svejgaard in Denmark)

HLA-A		HLA-B	
Allele	Percent	Allele	Percent
2	32.3	7	14.2
1	16.8	12	13.6
3	14.5	8	12.7
19	9.9	40	9.5
9	9.2	15	9.4
11	5.2	35	7.2
28	5.1	5	5.4
10	4.9	27	4.4
five or more other alleles	2.1	fourteen or more other alleles	23.6

Association of HLA with Autoimmune Diseases

The association of HLA type and degenerative autoimmune diseases came to light as a result of an experience on the USS *Little Rock* while it was on duty in the Mediterranean. After a shore leave, one of the cooks developed severe diarrhea. Shortly thereafter 602 of the 1,276 sailors on the ship had the same condition. The illness was caused by the bacterium, *Shigella dysenteriae,* which had been spread from the cook by careless food handling. Several weeks later, 10 of the sailors who had had the dysentery developed a painful, inflamed spine that persisted, and they had to be discharged. When the HLA types were discovered, it was found that 5 out of the 6 who could be located had HLA-B27 and had poker spine. The sixth one, who had another type, had a much milder reaction. Continuing study showed that over 90 percent of the persons who develop ankylosing spondylitis are of the HLA-B27 type. Hence, it is evident that those with this type respond to *Shigella* infection by the production of antibodies that often cause the spinal inflammation and eventual fusion of vertebrae.

A similar association has been found in some of the other autoimmune diseases; in fact, it is possible to predict the kind of autoimmune disease a person might develop by the HLA type. This relationship is shown in Table 7–2. Since the frequencies of the HLA types vary in different populations, the kind of autoimmune diseases varies in different populations. The particular types of HLA that react most strongly with the infectious agents in the environment will be selected and established at a high level. Hence, the people have a rather high degree of resistance to the infectious diseases, but they pay for this with a greater chance of developing specific autoimmune diseases

TABLE 7–2   Association of diseases with HLA type

Disease	HLA Type	Increased Risk Over Average	Symptoms of the Disease
Addison's disease	BA8	4	deficiency of hormones from cortex of adrenal glands; brownish skin discoloration, low blood pressure, debility
ankylosing spondylitis (poker spine)	BW27	150	inflammation of spine with eventual calcification of cartilages resulting in fusion of spine
anterior uvelitis (acute)	BW27	30	inflammation of iris and ciliary body of the eye
celiac disease	BA8	10	inflammation of small intestine (chronic) due to allergy to gluten of wheat
diabetes mellitus (juvenile form)	B8, B15, DW3	4	beta cells of islands of Langerhans of pancreas damaged, insulin output diminished, high blood sugar resulting
Grave's disease	DLD8a	11	hyperactive thyroid gland, producing thyroid antibodies that stimulate thyroid activity
hepatitis	A1, B8	10	inflammation of liver
Hodgkin's disease	AW5, 15, 18	2	cancer of lymph nodes, which become inflamed with decreased T cell functions
leukemia, lymphocytic	A12	13	cancer of lymphocytes, T cells in excess in blood

TABLE 7–2 Association of diseases with HLA type—continued

Disease	HLA Type	Increased Risk Over Average	Symptoms of the Disease
multiple sclerosis	DLD7a	5	sclerotic spots appear on brain and spinal cord, resulting in muscular weakness and speech disturbances
psoriasis	A13	5	inflammation of skin, scales develop mainly on scalp, elbows, and knees
Reiter's syndrome	BW27	40	inflammation of spine, prostate gland, and parts of the eye
systemic lupus erythematosus (lupus)	BW15 (whites) BW5 (blacks)	15	varied symptoms, may affect skin, joints, heart, lungs, kidneys, or brain

associated with the inherited HLA types. A balance is struck between these opposing forces.

While HLA types are certainly of great importance in determining susceptibility to autoimmune diseases, other factors can be involved. A mutation of one gene or a chromosome aberration in a lymphocyte can render it capable of reacting to body tissue as if it were foreign. Also, some of the lymphocytes that might react with body antigens may remain isolated in some pocket of the embryo and not be inactivated by contact with body cells. If these break out into the blood stream after birth they could produce antibodies against certain body tissues. Some body cells might also be so isolated during embryonic life that they are not contacted by the circulating B lymphocytes that later will not recognize these cells as self. We may inherit antigens in certain body tissues that are quite similar to those of certain microorganisms. Antibodies produced against these microorganisms may react with these body tissues. This may be the reason why our inherited HLA type makes us especially susceptible to certain autoimmune diseases. Remember that the HLA antigens are also found on many body tissues as well as on the leukocytes.

Treatment of Autoimmune Diseases

Autoimmune diseases are very difficult to treat because we are dealing with a natural body reaction that is necessary to protect us against infections. We can easily depress the immune system with immuno-suppressant drugs and that will relieve the autoimmune reaction, but at the same time will leave us open to infections that might easily be fatal. We must strike some compromise between the two extremes. For instance, myasthenia gravis has been greatly relieved by removal of the blood, a little at a time, and the extraction of gamma globulin from it. Some people who could not raise their arms from the bed have been able to be up and about after such treatment. Other treatments involve removal of the lymphocytes from the blood. This can be done by grad-ual removal of the blood and lowering the temperature to just below freezing. This will destroy the lymphocytes, but not the other blood cells. Then the blood is warmed and returned to the patient. Antihis-tamines may give some relief. Much of the irritation that results from antigen-antibody reactions is a result of the release of histamines from the cells and antihistamines counteract these.

Vaccination against agents that might lead to autoimmune reactions is also a possibility. If you are HLA type B-27, for instance, you might be immunized to *Shigella dysenteriae.* We must be careful, however, that the vaccination does not bring on the autoimmune reaction. Out of millions of people who received swine flu vaccine in 1977, about four hundred developed a paralysis and eight of these died. This relatively small group had HLA types that made the reaction possible. We could even determine the HLA type of a fetus by amniocentesis and know the particular autoimmune diseases that might be likely in the future. Some parents might even consider abortion if the fetus proved to have an HLA type that would most likely lead to serious debilitating dis-eases. An important moral issue is, of course, involved here.

PROBLEMS AND QUESTIONS

1   Those people who show a high resistance to infectious diseases seem to be the ones most likely to develop autoimmune diseases and vice versa. Give a possible explanation.

2   We have very efficient vaccines against smallpox, polio, and teta-nus, but no effective vaccine against the common cold. Give a pos-sible explanation for this difference.

3   What difficulties might arise if the fetus could respond to foreign antigens by the development of antibodies?

4 A man receives a kidney transplant and his doctor warns him that he is now more susceptible to cancer and should be on the lookout for early indications of possible cancer. Why would the doctor make this statement?

5 A man in England claims to have a son produced by cloning from a single cell from his body. If this were true, could this son give a kidney to his father without the need for immunosuppressants? Could the father give to the son? Give reasons for your answers.

6 The natives of a population in central Africa suffer from autoimmune diseases that are different from those found in natives of England. Give possible reasons for this difference.

7 Some women have antibodies against some HLA antigens even though they have never received a blood transfusion. How is this possible?

8 Most autoimmune diseases are characterized by irritation of a particular type of tissue, yet lupus (SLE) may affect many different kinds of tissue. Explain.

9 Heavy exposure to radiation or other mutagenic agents may be followed by the development of an autoimmune disease. How might this be correlated with the possible causes of autoimmune diseases?

REFERENCES

BURNET, F. M., Editor. 1976. *Immunology: Readings from Scientific American.* W. H. Freeman and Co. San Francisco.

CUNNINGHAM, B. A. 1977. The structure and function of histocompatibility antigens. *Scientific American* 237(4):96–107.

FUDENBERG, H. H., J. R. L. PINK, A.-C. WANG, and S. D. DOUGLAS. 1978. *Basic Immunogenetics,* 2d ed. Oxford University Press. London.

GOWANS, J. L. 1977. *Cellular Immunology.* Oxford/Carolina Biology Reader. Oxford University Press. London.

KOFFLER, D. 1980. Systemic lupus erythematosus. *Scientific American* 243(1):52–61.

LEWIN, R. 1974. *In Defense of the Body.* Anchor Press/Doubleday. Garden City, New York.

McMICHAEL, A., and H. McDEVITT. 1977. The association between the HLA system and disease. *Progress in Medical Genetics, New Series* 2:39–100.

NOSSAL, G. J. V. 1975. *Medical Science and Human Goals.* Edward Arnold. London.

NOTKINS, A. L. 1979. The causes of diabetes. *Scientific American* 241(5):62–73.

PORTER, R. R. 1976. *Chemical Aspects of Immunology.* Oxford/Carolina Biology Reader. Oxford University Press. London.

ROITT, I. M. 1977. *Essential Immunology,* 3d ed. Blackwell Scientific Publishers. Oxford.

ROSE, N. R. 1981. Autoimmune diseases. *Scientific American* 244(2):80–103.

SVEJGAARD, A., M. HAUGE, C. JERSILD, P. PLATZ, L. P. RYDER, L. S. NIELSEN, and M. THOMSEN. 1975. *The HLA System: An Introductory Survey. Monographs in Human Genetics,* Vol. 7. S. Karger. Basel.

Blood Genetics

HUMAN BLOOD IS AN EXCELLENT SUBJECT for genetic investigation for several reasons. *First,* it is one of the most variable parts of the human body, and variety is necessary for characteristics to be analyzed genetically. Blood is so variable that you could search the world over and never find two people with exactly the same combination of inherited blood traits. *Second,* blood shows comparatively little variation as a result of environmental influences. You inherit genes for body stature, yet your stature can vary considerably as a result of diet, disease, and other environmental factors. On the other hand, if you inherit a gene for type A blood, you have this type when you are born and will have it without alteration all of your life, no matter what environmental forces come to bear upon you. *Third,* blood is readily available and can be analyzed in the laboratory by immunological and electrophoretic techniques to give positive identification of its many unique features. *Fourth,* most variations in blood characteristics are the result of variations at single gene loci. For instance, variations of a single gene determine if you have type A or type B blood, but polygenes are involved in variations of many other human characteristics, especially externally visible characteristics.

In this chapter we shall learn something about inherited blood characteristics and their significance to us as individuals and to geneticists as they try to probe ever deeper into the secrets of human heredity. In the first part of the chapter we shall be concerned primarily with the antigens on the red blood cells and in the second part we shall consider the proteins in the plasma.

Red Blood Cell Antigens

We have been able to identify many antigens on the surface membranes of the erythrocytes (red blood cells) by mixing the cells with sera containing various antibodies. The best known of these antigens are related to the ABO blood groups or types.

The ABO Blood Groups

Discovery of Blood Groups   Attempts at blood transfusion were made in the eighteenth century. Some of these were successful, but many of the recipients died shortly after the foreign blood entered their veins. It was not until 1901 that the reason for these reactions was discovered. Karl Landsteiner separated blood cells from plasma with a centrifuge and then made various recombinations. The defibrinated plasma, or **serum,** from one person would sometimes mix smoothly with cells from other persons, but in other cases the cells would clump together in large aggregations (Figure 8–1). This clumping is known as **agglutination.** The reactions could be explained on the basis of two antigens, which we now call A and B. All people can be classified into four blood groups according to these antigens. All people also have antibodies against all of the antigens possible except those that would react with and agglutinate their own red blood cells. This means that those with type A blood have the A antigen. They could not have anti-A antibodies, for these would clump their own cells, but they can and do have anti-B antibodies. Figure 8–2 shows the antigens and antibodies in the four blood types.

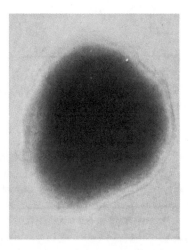 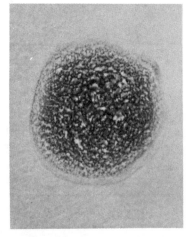

FIGURE 8–1.   Blood from a person with type A has been mixed with serum from a person with type B at right. This serum contains anti-A antibodies and has agglutinated the red blood cells. At left, some of the same blood has been mixed with serum from another type-A person, and there is no agglutination.

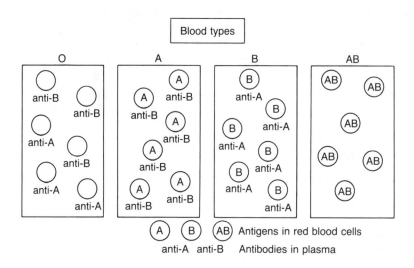

FIGURE 8–2. Antigens and antibodies found in the four blood types.

Blood Typing and Transfusion   In transfusions, only blood that does not contain antigens on the red blood cells that will be clumped by the antibodies of the recipient can be given. Two kinds of serum are used in blood-typing. One contains anti-A antibodies and is the serum from a person who is type B. The other contains anti-B and is the serum from a person who is type A. A person whose blood cells are not agglutinated by either of these sera is type O. One whose cells are agglutinated by anti-A, but not by anti-B, is type A. One whose cells are agglutinated by anti-B, but not by anti-A, is type B. Finally, one whose cells are agglutinated by both types of sera would have both A and B antigens and would be type AB.

As a rule, a person is given transfusions only from a person of the same blood type, but in an emergency some combinations can be made. The important thing to remember is that blood that will be clumped by antibodies of the recipient is never given. Agglutination of the red blood cells in the blood vessels of the recipient will cause obstruction of the capillaries, usually resulting in death.

In certain cases in years gone by, blood was given that contained antibodies against the cells of the recipient. Type O blood was given to type A individuals even though the O plasma contains anti-A. This is possible because the antibodies in the donor plasma are quickly diffused through the blood of the recipient and the titer becomes too low to cause much of a reaction. Blood from type A, however, cannot be given to type O because the high titer of anti-A in the recipient would quickly clump the A cells as they enter.

Inheritance of the ABO Blood Groups A set of three multiple alleles at an autosomal locus is responsible for the four blood types. One of these, which is recessive to the other two, produces neither antigen. Another allele produces antigen A and is codominant with the third allele, which produces antigen B. The gene symbols i or I^O, I^A, and I^B are often used, with the I standing for isohemagglutinin and the superscripts indicating the antigen produced. Another gene involved in blood characteristics, however, is designated as I, so we shall use another system, which also has the advantage of indicating dominance relationships:

a—neither antigen A—A antigen A^B—B antigen

These three alleles can be combined to give six genotypes and four phenotypes:

Genotypes	Phenotypes (blood types)
a/a	O
A/A or A/a	A
A^B/A^B or A^B/a	B
A/A^B	AB

This explains how the antigens are inherited (see also Figure 8–3), but what about the antibodies? Most antibodies are produced in response to exposure to antigens, but anti-A and anti-B seem to arise

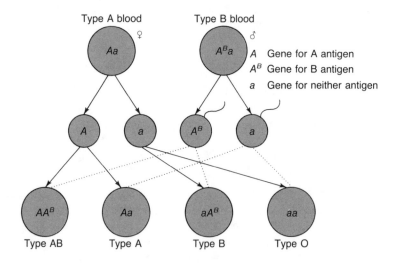

FIGURE 8–3. Blood types to be expected in the children of a man who is type A and a woman who is type B when both also carry the recessive allele for type O.

without any exposure to A or B antigens. One theory suggests that human embryos produce B lymphocytes programmed for the production of anti-A and anti-B antibodies, just as they produce many other preprogrammed B lymphocytes according to the clonal selection theory. No antibodies are produced before birth, but the contact of these B lymphocytes with the antigen to which they are specific will inactivate them. Hence, in those with the A antigen, the B lymphocytes for anti-A are inactivated. If the B antigen is present, it will inactivate the B lymphocytes for anti-B.

About two to eight months after birth, the B lymphocytes that have not been inactivated begin producing plasma cells that will release antibodies, even though there is no apparent contact with the antigens. A type O person has inactivated neither of the two kinds of B lymphocytes, so produces both types of antibodies; a type A person has inactivated those for anti-A and produces only anti-B; and so on. One part of the theory holds that no contact with a foreign antigen is necessary to start antibody production, that these two types of B lymphocytes need no collaboration with T lymphocytes or macrophages. If this is true, they are unique in this regard because no other cases of spontaneous production of antibodies are known. An alternative theory holds that these two are not any different from other antibodies and are actually produced after stimulation by a foreign antigen. According to this theory, antigens very similar to A and B are prevalent in foods, bacteria, and other substances that may enter the body, and contact with these antigens stimulates the immune reaction and anti-A and anti-B are produced. Support for this latter concept has been provided by experiments in which a small amount of A antigen is injected into persons who are type O or B. A very rapid increase in the titer of anti-A results, much faster than would be expected if this were the first contact with antigens of this nature. It is similar to the reaction to a booster shot after a vaccination, when "memory cells" are stimulated into production of more plasma cells and more antibodies.

Legal Applications   Since the inheritance of the ABO groups is so definite, blood typing is often used to determine non-parentage when disputes arise. Say, for instance, that a woman sues a man for support of her child, claiming that he is the father, but that the man denies the claim. By typing the mother and child it becomes evident that the biological father had to carry certain genes. Therefore, if the blood type of the purported father shows that he does not carry these genes, then he can be ruled out as the biological father. Of course, in many cases a man will have the type required, but this does not prove that he is the father, because other men are of this type also. However, as we have learned about other antigens in the blood, further tests usually can be definitive. Divorce suits claiming infidelity on the part of the mother and inheritance claims from long-lost relatives are other examples where blood genetics can help settle legal questions related to inheritance.

Precursor of A and B Antigens   In Bombay, India, a man was found who had blood that was not agglutinated with either anti-A or anti-B serum, yet his parents were type A and AB. A type O child from such parents would seem impossible. Furthermore, he was married to a woman who was type O and had children with type A blood, again a seemingly impossible occurrence. He must have carried the gene for A, but did not produce the A antigen. Tests of his blood showed that he lacked **substance H,** a precursor from which the A antigen is produced. He proved to be homozygous for a recessive gene, *h,* which prevented him from producing substance H. Both A and B antigens can be produced from this precursor. Most people carry at least one dominant allele, *H,* and produce the precursor. This man, who appeared to be type O, was said to have the **Bombay phenotype.** Others with this phenotype have been found in many other parts of the world, yet they are comparatively few. The Bombay phenotype illustrates another case of epistasis; *h* is epistatic to both *A* and A^B.

The H substance can be identified because red blood cells containing it are agglutinated when they are mixed with serum from a certain species of eel or an extract of the plant *Ulex europeus* (Figure 8–4). Those with type O blood have an abundance of this substance because

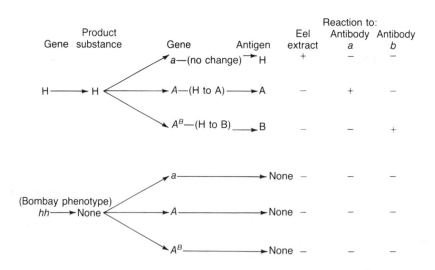

FIGURE 8–4. Relation between the precursor H substance and the A and B antigens. Those with the Bombay phenotype cannot produce substance H and, hence, cannot have either A or B antigens even though they may have genes for these antigens.

they have the gene *H* that produces it, but they do not use it in antigen production. Those heterozygous *A/a* or *A^B/a* have some H substance because they do not use all of it. However, those homozygous for *A* or *A^B*, or who carry both of these genes, seem to use practically all the H substance in antigen production, so their red blood cells are not agglutinated.

This discovery has made it possible to narrow down potential relatives in legal disputes. As an example, a woman who is type O and has a baby who is also type O sues a man for child support. He proves to be type A and could be the biological father if he is heterozygous *A/a,* but not if he is homozygous *A/A.* A test for the H substance in his blood will show which he is.

The Secretor Trait

Some people with A or B antigens in their blood cells also have these antigens in their body secretions, such as saliva, gastric juice, and secretions of the nose and eyes. Those who are type O, or heterozygous for A and B, will also have the H substance in their secretions. Such people are known as **secretors,** and those who do not have the trait are **nonsecretors.** A dominant gene, *Se,* codes the secretor trait. It can be identified as follows. Let us assume that a man has type A blood. A drop of his saliva is mixed with a drop of anti-A. No reaction can be observed, but the antigen in the saliva combines with the antibodies in the serum and binds them. Then a drop of the man's blood is added to the mixture. If he is a secretor there will be no reaction because the antibodies, bound to the antigens in the saliva, are no longer free to react with the antigens on the blood cells. If he is a nonsecretor, however, the red blood cells will be agglutinated because the anti-A antibodies are not bound since there was no antigen-A in the saliva. Type B secretors can be recognized by mixing saliva with anti-B. Even type O secretors can be recognized by using anti-H to bind the H substance, which will be in the saliva of secretors.

The discovery adds another gene difference that can be used to settle cases of disputed parentage or in crime laboratories to help identify criminals. A kidnapper may write a letter demanding ransom. An analysis can be made of the dried saliva used to seal the envelope to show if it was sent by a secretor or nonsecretor. The type of blood can even be determined if the person was a secretor. Blood tests of an arrested suspect will help identify the sender of the letter. Semen obtained from rape victims can also be used in this manner to help identify the offender.

The Rh Antigens

Blood transfusions began to be used widely after the ABO groups were discovered, but a few cases were puzzling. Sometimes agglutination would occur even though both donor and recipient were of the same type. The reason for this became apparent when the **Rh factor** was discovered. In 1940 Karl Landsteiner found a factor in the blood of the Rhesus monkey that was also present in the blood of some people. He discovered it by injecting red blood cells from monkeys into rabbits. The rabbits produced antibodies against some antigen in these cells. Serum from such rabbits would agglutinate the red blood cells of all Rhesus monkeys as well as those of some people. Could it be possible that some humans shared the gene for the production of this antigen with the monkeys? Indeed they could, and did. The antigen causing the reaction is known as the **Rh factor** and a dominant gene seemed to produce it. People with the antigen are known as **Rh-positive** and those without it are **Rh-negative.**

Human Reactions to the Rh Factor   Humans do not produce the Rh antibodies naturally, as they do the anti-A and anti-B, but those who are Rh-negative can produce anti-Rh if they receive blood containing the Rh factor. This explains some of the early difficulties with transfusions. The first transfusion of positive blood into a negative recipient was successful if the ABO types matched, but the Rh antigens in this blood would stimulate the production of anti-Rh. Therefore, future transfusions of Rh-positive blood would result in agglutination of the donor's cells. Hence, blood must be tested for the Rh factor as well as for the ABO type before transfusion. Persons who are type O, Rh-negative are limited in the donors from which they can receive transfusions. About 47 percent of the persons in the United States are type O, but only about 15 percent are Rh-negative. This reduces the options to about 7 percent of the potential donors ($0.47 \times 0.15 = 0.07$).

Rh Incompatibility and Childbirth   Another medical mystery was solved when the Rh factor was discovered. Many babies were born with **erythroblastosis fetalis,** also called **hemolytic disease of the newborn (HDN).** These babies were anemic, jaundiced, and had deformed and nucleated red blood cells. They often died before birth or within a few days after birth, but if they survived for a week or so they gradually replaced their defective cells with normal cells. The condition almost never appeared at a first birth, but became more common at each succeeding birth (Figure 8–5).

When the Rh factor was discovered, it was found that Rh-positive babies with this defect were born to mothers who were Rh-negative, and only after the mothers had already borne a previous Rh-positive baby. During birth, as the placenta separates from the uterus, considerable bleeding occurs and many of the fetal blood cells enter the mother's blood stream. It is as if she receives a small transfusion from her baby. An Rh-negative woman can thus be sensitized to the Rh antigen

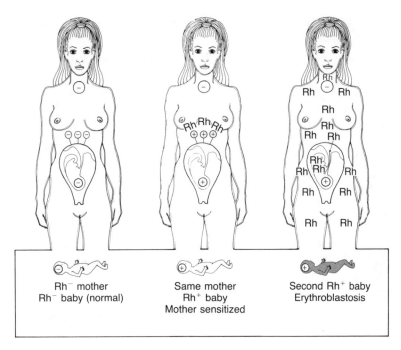

Rh⁻ mother
Rh⁻ baby (normal)

Same mother
Rh⁺ baby
Mother sensitized

Second Rh⁺ baby
Erythroblastosis

FIGURE 8–5. An Rh-negative woman can be sensitized when she bears an Rh-positive child, and future Rh-positive children may have erythroblastosis. In this illustration, the third child is afflicted.

if it is present in these fetal cells. When she bears a second child who is Rh-positive, some of the anti-Rh in her blood can pass through the placenta and react with the Rh antigen in the cells of the fetus. The titer of antibodies is not sufficient to cause agglutination, but may be high enough to destroy and damage many of the fetal cells. Anemia results from the loss of many red blood cells. To compensate for this loss, the body releases many immature, nucleated cells, which cannot carry oxygen as well as normal cells (Figure 8–6). Jaundice, another complication of HDN, results from the accumulation of bilirubin, a toxic pigment resulting from the breakdown of damaged red blood cells. Bilirubin can cause brain damage, mental retardation, and cerebral palsy. Blood transfusions immediately after birth can save many afflicted babies.

In the past, Rh-negative women sometimes tried to find Rh-negative husbands so that they would not have to worry about Rh sensitization. Today, however, thanks to a discovery made in the 1960s, this is no longer necessary. In fact, Rh-induced erythroblastosis is declining in frequency because of the widespread use of this technique. Whenever a negative woman bears a positive child she now is given an injection of

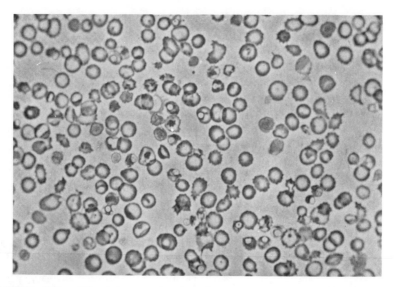

FIGURE 8–6. Blood from a baby with erythroblastosis. Many of the red blood cells have nuclei because they have been prematurely released from the bone marrow.

rhogam, the gamma globulin fraction of blood from a negative person who has been sensitized to the Rh antigen (Figure 8–7). It contains anti-Rh antibodies that react with and bind the Rh antigens in fetal cells that have entered her blood. As a result, there are no free antigens to sensitize her and she can bear as many positive babies as she pleases, as long as she receives rhogam after each birth. Rhogam must also be

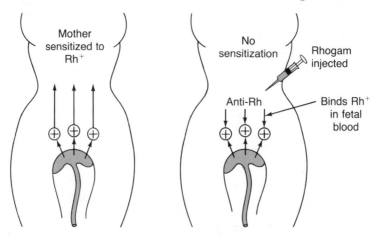

FIGURE 8–7. Injection of anti-Rh (rhogam) into an Rh-negative woman after the birth of each Rh-positive baby can prevent sensitization of the woman by binding Rh-positive red blood cells from the baby.

given after an abortion or miscarriage because some blood from the embryo may enter the mother's body. Rhogam has no value for a woman who is already sensitized, however, because she already has the antibodies.

Rhogam is obtained from men or postmenopausal women who are Rh-negative. They are sensitized by injection of the antigen that has been removed from the blood cells of a positive person. Then a pint of blood is removed and the gamma globulin fraction extracted. This fraction contains the antibodies. The rest of the blood is then returned to the donor so only a small amount of blood is lost. This procedure can be repeated over and over again, with occasional booster shots of Rh antigen to keep the antibody titer high.

Complexities of the Rh Factor   Continued investigations of the Rh factor have shown that it is much more complex than was at first recognized. A number of different antigens can be produced. To explain this, A. S. Wiener proposed a theory that a single gene locus is involved, but a great variety of alleles may be possible at this locus. Furthermore, each of these alleles can produce from none to three antigens. An alternative theory, proposed by Fisher and Race, holds that three very closely linked gene loci are involved with two major alleles possible at each locus. Since this latter system is easiest to understand and most widely used, we shall adopt it for our discussions. The alleles at the three loci are designated as *C, c; D, d;* and *E, e.* All of these six alleles, except *d,* produce antigens that can be detected by antibody reactions. Of the five antigens that can be produced, however, only antigen D is strong enough to cause difficulties in transfusion and childbirth in most cases. In a few cases the antigen c causes some difficulty. Very rarely the other three cause a mild reaction, so it is customary to test only for D and sometimes for c in cases of transfusions and childbirth. As a rule, only those who produce antigen D are considered Rh-positive; all other combinations are Rh-negative.

The most common allelic complex is *CDe,* which gives rise to three antigens, but since each person is diploid, as many as five possible antigens can be present in one person. The three most common diploid genotypes among North American whites are

Genotype	Frequency	Antigens Present
CDe/cde	0.33	C,D,c,e
CDe/CDe	0.17	C,D,e
cde/cde	0.15	c, e

As is true for the A and B antigens, a precursor substance seems to be necessary for the Rh antigens. This concept is supported by the fact that a few persons have been found who are Rh^{null}. They have no Rh antigens on their blood cells even though they have inherited genes for some of these antigens.

Fetal Reaction to Maternal ABO Antibodies   Since anti-Rh antibodies in a mother's blood can so seriously affect an Rh-positive fetus, you may wonder why maternal antibodies of the ABO groups do not also cause reactions. Actually they do, but not in all cases of maternal-fetal incompatibility. One reason seems to be that anti-A and anti-B molecules are of the IgM type and are about 2.5 times larger than the IgG molecules of the Rh antibodies. Hence, anti-A and anti-B do not pass through the placental barrier as readily as the anti-Rh. Sometimes, however, some do get through and cause a severe reaction early in fetal life, during the first seven or eight weeks, which is much earlier than reactions to anti-Rh appear. The result is usually fetal death and abortion. Some authorities estimate that the ABO reactions are a major cause of fetal death, accounting for the loss of about 2 percent of all embryos conceived. Statistical studies of children from type A mothers and type O fathers showed more children with type A blood than in the reciprocal type of pairing. This statistic agrees with the expectation that some of the type A fetuses in type O mothers are lost because they receive anti-A from the mother.

Further evidence is provided from studies of the percentage of erythroblastosis from Rh incompatibility in different ABO type parents. An Rh-negative mother who is type O is more likely to escape sensitization to the Rh antigen from a fetus who is Rh-positive, type A rather than one who is Rh-positive, type O. Apparently, when cells from the type A fetus enter the mother's body they are destroyed by her anti-A and removed from the blood stream before the Rh antigen can sensitize her. When the mother is ABO compatible, however, the Rh-positive fetal cells remain and she becomes sensitized.

Other Red Blood Cell Antigens

Continued improvements in the techniques of immunogenetics have led to the discovery of quite a number of other antigens on the red blood cells. We shall mention some of the better known antigens.

MN and Ss Antigens   As more was learned of blood antigens, research workers suggested that some antigens might be on human cells for which the human system is incapable of producing antibodies. To test this idea, some human red blood cells were injected into rabbits, and serum from these rabbits was later mixed with blood from different persons. It was found that the serum would clump cells from some persons and not those from others. Two new antigens, M and N, were identified in this way. People can be classified according to these antigens as M, N, and MN. (No equivalent of O in the ABO series was found.) Humans do not seem to have B lymphocytes that can produce anti-M or anti-N, so these antigens present no problem in transfusions or childbirth. They are, however, very useful in legal and population studies.

More recently, two additional antigens, designated S and s, were found. These seemed to be produced by a gene locus so close to that for M and N that the genes tend to remain as a unit in meiosis. Hence, we classify people according to the two antigens as MS, Ms, MNS, MNs, NS, and Ns. The gene symbols—L^{MS}, L^{Ms}, L^{NS}, and L^{Ns} (the L stands for Landsteiner, who discovered the system)—suggested by some workers imply that a single gene locus controls the M, N, S, and s antigens.

Additional Antigens   Still other antigens have been discovered on the red blood cells of children who have been born with mild erythroblastosis when there was no Rh or ABO incompatibility. In Boston, Mrs. Kidd had such a child at her second delivery and, in searching for the reason, it was found that her husband's blood contained a previously unknown antigen that she did not have. It was designated as Jk(a$^+$) and was produced by the gene Jk^a, for which he was homozygous. She was homozygous for an allele, Jk^b, and produced the antigen Jk(b$^+$). Both her children were heterozygous and, since the alleles are codominant, they produced both antigens. Because she had been sensitized to the Jk(a$^+$) antigen at the first birth, her antibodies reacted with the blood cells of her second child.

A very rare recessive allele, *jk,* has been found that gives the phenotype Jk(a$^-$b$^-$) when homozygous. This is known as the **Kidd system.** The antigens of this series are of minor clinical importance because so few people can produce antibodies against them and, when they do, the reaction is weak.

Still other antigens have been discovered when patients had mild reactions to repeated transfusions from the same donor or when fetal blood was damaged even though there was no apparent incompatibility between the bloods involved. Space limitations prevent a complete account of all of these, but here are the names of some of the systems: **Auberger, Diego, Dombrock, Duffy, I, Kell-Cellano, Lewis, Lutheran,** and **Xg**. The Diego system is interesting because all American Indians and about 10 percent of Japanese people are Diego-positive, but virtually all Caucasians are Diego-negative. The Xg system is valuable because the gene locus is on the X chromosome near the end of the short arm. It can, therefore, be used as a marker for studies of other X-linked genes.

The two antigens of the Duffy system are significant because they serve as receptors for the malarial parasite, which easily becomes attached to the red blood cell membranes when these antigens are present. Those who inherit the recessive allele in a homozygous state have neither of these antigens, have no such receptors, and thus are highly resistant to malaria. Hence, selection would favor those who are Duffy negative in regions where malaria is prevalent. Indeed, we do find a high incidence of the allele for Duffy negative in people who live in regions of central or western Africa, or whose ancestors came from these regions. Malaria has been common in these parts of Africa for a long time. On the other hand, the Duffy negative allele is very rare in people from areas where malaria has not been a major problem.

Proteins Within Red Blood Cells

In addition to the many variations of the antigens on the surface membranes of red blood cells there is also great variation in the proteins within these cells. We shall mention two of them as examples.

Hemoglobin Variations

Hemoglobin is the major protein in red blood cells. The different kinds of hemoglobins, however, are not generally recognized by antigen-antibody reactions and were not understood until the technique of electrophoresis was discovered. The principle of this technique is described later in this chapter; suffice to say here that it enables us to identify hemoglobins that differ by only one amino acid in the long polypeptide chains that make up the molecule. These variations are such a good illustration of the principle of mutation that a complete discussion of them will be deferred until Chapter 10.

Erythrocyte Acid Phosphatase

Acid phosphatase is another protein found within red blood cells. It is an enzyme that is particularly interesting to geneticists because a measure of the amount of enzyme activity in a person's cells gives an accurate indication of the genotype of that person. There are three alleles, each of which is contributing; that is, each produces a specific amount of enzyme. The allele P^A contributes about 62 units, P^B contributes about 94 units, and P^C contributes about 118 units. If we find a person with an enzyme activity of about 156 units, we could assume that the person was heterozygous P^A/P^B. No other combination would give an activity of about this quantity.

Leukocyte Antigens

The leukocytes also have antigens on their cell surface membranes; these antigens have significance primarily in organ transplants. They are known as HLA (human leukocyte antigens) and are also important in autoimmune diseases. Since these have already been discussed in

Chapter 7, we shall only mention them here for the sake of complete-
ness in our discussion of the topic of blood genetics. At least four different gene loci are involved in the HLA antigenic variations. Hence, a
person heterozygous for the multiple alleles at each of these loci could
have eight different antigens on the leukocytes. Consequently, the
great variety of phenotypes that exists can be valuable in studies of
populations and in genetic-legal analyses.

The Plasma Proteins

In addition to the great variety in human blood based on the proteins
associated with the blood cells, additional diversity occurs among the
plasma proteins. Three of the plasma proteins are known as alpha,
beta, and gamma globulins. In addition, there is **fibrinogen,** which
forms blood clots; **complement,** which is involved in some immunological reactions; and **albumin,** which is the most abundant of the blood
proteins and gives blood its viscid consistency. We can also include
various enzymes that are circulating in the plasma. Gamma globulin
was considered in Chapter 7.

The plasma proteins are usually identified by electrophoretic separation of their component parts. Such separation is possible because
different proteins have different electrical charges that cause them to
move at different speeds in an electrical field. The variation in charges
is due to different proportions of positively and negatively charged
amino acids in the protein side chains. Of the twenty amino acids that
may be a part of the polypeptide chains of proteins, two have an overall
negative charge and three have an overall positive charge; the others
are neutral. Since the amino acid composition of proteins varies, most
protein molecules will be charged to different degrees.

Entire proteins can be separated from one another by the use of a
starch or acrylamide gel. A drop of the proteins can be placed at a
certain spot on this gel and then exposed to a strong electrical field.
Proteins with the strongest negative charges will migrate farthest
toward the positive pole in a given time and vice versa (Figure 8–8).
The gel strip is then treated with a dye that stains the proteins and
makes them visible as distinct bands.

If the proteins are treated with a digesting enzyme, they will be broken up into small peptide chains and each of these will have a different
speed of migration in an electrical field. We can go further and break
up the peptides and determine the specific amino acids in the chains.
Since genes determine the amino acid composition of proteins, variations in amino acid sequence indicate gene differences. Let us now see
what electrophoresis has done to give us information on the plasma
proteins.

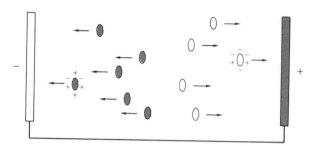

FIGURE 8–8. Protein molecules will move in an electrical field. Those that are more negatively charged will move toward the positive pole, while those that are more positively charged will move toward the negative pole. Using electrophoresis, the various proteins present in a mixture can be separated.

Haptoglobins (Alpha Globulins)

The **haptoglobins** bind free hemoglobin that is released when red blood cells burst. They transport it to the liver where it is broken down, and the iron recovered is for use in producing more red blood cells. This is an important function because free hemoglobin in the blood is toxic. Different varieties of haptoglobins, when they are separated into their component amino acids through electrophoresis, can be identified. Two codominant alleles, Hp^1 and Hp^2, give rise to three basic haptoglobin types: Hp 1, Hp 2, and Hp 1–2. A third allele, which is of low frequency, produces a nonfunctional haptoglobin and those homozygous for this allele suffer from the toxic effects of free hemoglobin in the blood. While this gene is rare in most parts of the world, it is fairly common in parts of equatorial Africa. Small variations of each of the codominant alleles give as many as eighteen recognizable kinds of haptoglobins. Two amino acid chains actually make up the haptoglobin molecule. They are known as alpha and beta chains. The variations described are all in the alpha chain; the beta chains seem to be the same in all people. Hence, all people would seem to be homozygous for the same gene that produces the beta chain.

Transferrins (Beta Globulins)

From the intestine, where it is absorbed from food, ionic iron is transported by transferrins to the body parts where it is used. A transporting agent is necessary because free ions of iron in the blood are toxic. The most common type is **transferrin C**, coded by the gene Tf^C. Two basic alleles of this gene have been identified by electrophoresis. One, Tf^Bm, codes for a globulin more negatively charged than transferrin C, and the other, Tf^D, codes for a more positively charged globulin. Variations of these two, however, push the total number of alleles beyond twenty. All appear to be codominant, but one recessive allele,

Tf^O, has been found that fails to produce functional transferrin. People homozygous for this gene have **atransferrinemia** and usually die young because of the toxic effects of free iron ions in the blood.

Fibrinogen

Fibrinogen is the plasma protein that solidifies into fibrin when blood clots. A recessive gene fails to produce this protein and the result is **afibrinogenemia,** in which the blood cannot clot. Bleeding in persons with this defect will eventually stop if direct pressure is applied because the walls of the injured blood vessels will adhere if the flow is not too great.

Alpha antitrypsin

This is an enzyme in the albumin complex, and it shows many variations that have significance in certain diseases. The enzyme neutralizes trypsin and other proteolytic enzymes, thus preventing damage to cells. Over twenty-three different forms of this protein have been identified by electrophoresis. Multiple alleles at a single locus are responsible. The basic gene symbol is Pi and the different alleles are indicated by superscripts, as Pi^M, Pi^Z, etc. The phenotypes that result are known as MM, MZ, etc. The ZZ phenotype is especially important because it causes a decided reduction in the activity of antitrypsin. The low activity of the enzyme associated with the ZZ phenotype appears to be related to pulmonary emphysema. This disease, which blocks off large areas of the lungs from active usage, is about fifteen times more frequent in people whose genotype is ZZ as compared to those with other alleles. People who are heterozygous for the Pi^Z allele have a much lower frequency of emphysema, but it is still higher than for those who do not have this allele.

These findings are especially important as they relate to cigarette smoking and emphysema. We know that some people smoke heavily for many years without developing emphysema, while others develop it after only a few years of heavy smoking. The difference may be due to their relative activities of alpha antitrypsin. Smoking causes lung inflammation during which proteolytic enzymes are released to counter the effect of the irritation. If they are not sufficiently inhibited by the alpha antitrypsin, however, the proteolytic enzymes may do damage to the cells of the lungs so the cells can no longer function. Those who inherit a low antitrypsin production might well be advised to avoid smoking and working in areas where lung irritants might be inhaled. Also, every effort should be made to avoid infection, such as bronchitis, that would cause lung irritation.

Those with reduced antitrypsin activity are also more susceptible to cirrhosis of the liver, especially during childhood. This degeneration of liver tissue is not as common as emphysema, however, possibly because some precipitating factor in the environment is not as common as lung irritants.

PROBLEMS AND QUESTIONS

1   A woman who has type A blood marries a man with type B. They have four children and are amazed to find that each child has a blood type different from the other children. Give the genotypes and phenotypes of the parents and each of the four children.

2   A woman who has blood type A bears a child who has type O blood. The woman's husband has blood type B, and he feels that this proves that he is not the biological father of the child. Is he justified in this assumption? Use genotypes to illustrate your answer.

3   A couple brings a newborn baby home from the hospital, but feels that this baby does not look like the one the wife bore. They ask for a blood test, which shows that all three have blood type A, but further tests show that the man is MS, the woman is Ns, and the child is Ns. Do these results confirm their suspicions that a mix-up has occurred? Explain.

4   Would people with the Bombay phenotype be desirable as blood donors? as blood recipients? Explain your answers.

5   In the few cases where ABO incompatability causes hemolysis of fetal red blood cells, it has occurred just as often at the first birth as at succeeding births; but all other hemolysis caused by antigen incompatability occurs only after at least one birth or a transfusion. Explain.

6   A young lady who is about to be married learns that she was born with hemolysis resulting from Rh^D incompatability. She fears that the same condition might arise in her children. What could you tell her that might ease her concern?

7   How can the injection of anti-D, the very antibody that causes hemolytic disease of newborns, prevent such births in the future?

8   A banker receives an extortion letter, which is turned over to the FBI for analysis. The FBI tests the saliva used in sealing the envelope and finds that A antigens are present. What three genes do you know for sure were present in the sender of the letter?

9   A man who is a heavy smoker develops emphysema; tests show that he has very low antitrypsin activity in his blood albumin. His wife has a high activity of this enzyme. They have two children and are worried about their possible susceptibility to emphysema. Evaluate their chances according to their inherited antitrypsin activity.

10 Studies of the differences in the plasma proteins are usually made by electrophoresis, while studies of the protein differences on red blood cell membranes are usually done by antigen-antibody reactions. Why do you think these different methods are used?

11 A woman is found who has the very high erythrocyte acid phosphatase activity of 236. What would you expect her genotype to be?

REFERENCES

CLARKE, B. 1975. The causes of biological diversity. *Scientific American* 233(2):50–60.

CLARKE, C. A. 1972. Prevention of Rh isoimmunization. *Progress in Medical Genetics* 8:169–223.

EDWARDS, J. H. 1965. The meaning of the associations between blood groups and disease. *American Journal of Human Genetics* 29:77–78.

ERSKINE, A. G., and W. W. SOCHA. 1978. *The Principles and Practice of Blood Grouping,* 2d ed. The C. V. Mosby Co. St. Louis.

GIBLETT, E. R. 1977. Genetic polymorphisms in human blood. *Annual Review of Genetics* 11:13–28.

LANDSTEINER, K., and A. S. WIENER. 1940. An agglutinable factor in human blood recognized by immune sera for rhesus blood. *Proceedings of the Society for Experimental Biology* 43:223–46.

MOURANT, A. E., A. C. KOPÉC, and K. DOMANIEWSKA-SOBCZAK. 1976. *The Distribution of the Human Blood Groups and Other Polymorphisms,* 2d ed. Oxford University Press. London.

ORTHO DIAGNOSTICS. 1969. *Blood Group Antigens and Antibodies as Applied to the ABO and Rh Systems.* Ortho Diagnostics. Raritan, New Jersey.

RACE, R. R., and R. SANGER. 1975. *Blood Groups in Man,* 6th ed. Blackwell Scientific Publications. Oxford.

WHO SCIENTIFIC GROUP. 1971. *Prevention of Rh Sensitization.* World Health Organization Technical Report Series, No. 468.

Enzymes— The Agents of the Genes

WE OFTEN THINK OF ENZYMES IN TERMS OF the secretions that bring about changes in food in the digestive tract. These are extracellular enzymes, those produced by cells and then passed out to function elsewhere. Most enzymes, however, are intracellular. They remain within the cells where they are produced and bring about chemical changes within these cells. A different enzyme is required for each of the many thousands of reactions that take place in cells and a different gene must be present to code the production of each enzyme. If even one of these genes fails to produce its enzyme, or produces the enzyme in a defective form, serious alterations of body processes can result. Some of the most serious genetic diseases result from such alterations. Although some genes regulate the production of other protein products, in this chapter we will restrict discussion to gene-dependent enzyme deficiency diseases.

Enzyme Deficiency Diseases

Most genes that code the production of enzymes are dominant; one gene can produce all the enzyme required for normal function. Hence, a recessive allele that fails to produce a normal enzyme does no harm as long as the dominant allele is also present. Individuals homozygous for such recessive alleles, however, may suffer from an enzyme deficiency disease.

Galactosemia
A young couple had a baby who was a bright, physically normal boy. Several weeks after birth,

however, some alarming disorders began to appear. The baby became dull and showed little interest in his surroundings; his stomach became bloated and he would vomit up much of his milk after nursing. His liver also became enlarged and a cloudiness developed on the lenses of his eyes. The mother remembered an aunt who had a child with similar symptoms. That child continued to deteriorate, became severely retarded mentally, became blind because of cataracts, and had constant digestive upsets before it died at three years of age. These symptoms are caused by a deficiency of an enzyme needed in the processing of certain sugars. The condition, known as **galactosemia,** results when a child is homozygous for a recessive gene that fails to produce a liver enzyme necessary for the breakdown of one of the intermediate products of lactose or milk sugar (Figure 9–1). This sugar is broken down by intestinal enzymes into two simpler sugars, glucose and galactose. After these are absorbed, a liver enzyme attaches a phosphate onto the galactose and it becomes galactose phosphate. Another liver enzyme, **galactose phosphate uridyl transferase (GPT),** changes galactose phosphate into glucose phosphate; it is the enzyme GPT that is missing in individuals who have galactosemia. In galactosemia the concentration of galactose phosphate rises in the blood and causes the damage described.

Fetuses that lack GPT can function normally because the galactose conversion is made in the mother's liver, but once the umbilical cord is severed the babies become dependent on their own enzymes. Since milk

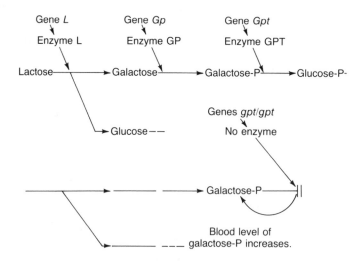

FIGURE 9–1. Galactosemia results when a person cannot produce an enzyme involved in the breakdown of lactose (milk sugar). At top the reactions are shown when the enzyme is present, and at bottom when the enzyme is missing.

is the first food ingested, the galactose phosphate buildup begins shortly after birth, although it will be several weeks before any symptoms are apparent. Such babies can be saved, however, by simply withholding milk from the diet, since milk is almost the sole source of lactose. A milk-free diet instituted soon after the symptoms appear will restore normal development; even the cloudiness of the lenses will clear up. This diet restriction must be continued throughout life; even foods prepared with milk must be excluded to prevent a return of the symptoms. The normal state cannot be restored, however, if the removal of milk from the diet is delayed very long after the symptoms have appeared.

Discovery of Enzyme Deficiency Diseases

Early in this century an English physician, A. E. Garrod, became interested in the cause of an unusual type of urine excreted by some of the babies he attended. This urine turned black when exposed to light and air. Parents became understandably concerned when they noticed black stains on the diapers of their babies. A chemical analysis of the urine showed that it was rich in **homogentisic acid,** also called **alkapton.** Garrod called the condition **alkaptonuria.** He reasoned that the baby was not producing some enzyme that normally breaks down the alkapton. The excess of alkapton does not do any apparent harm during early life, but at maturity it can cause some disfigurement because it accumulates in the cartilages, which then turn black where they are exposed to light. This includes cartilages in the ears (Figure 9–2) and at the tip of the nose, as well as in the sclera (white) of the eyes. The accumulation in the cartilages at the joints may also cause a painful type of arthritis.

Garrod also studied other biochemical defects in humans—notably albinism, cystinuria, and pentosuria. He recognized that each of these

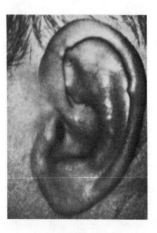

FIGURE 9–2. Ear of a man with alkaptonuria. Alkapton (homogentisic acid) accumulates in the cartilages and turns dark where it is exposed to light.

conditions had its origin in a single gene defect that led, ultimately, to interference with normal biochemical reactions. He published his conclusions in 1909 in a book called *Inborn Errors of Metabolism,* but the great significance of his findings was not realized until the 1950s, when the development of molecular genetics revealed the accuracy of his conclusions.

Enzymes in Series

Most of the enzyme-mediated reactions in cells occur in a series of small changes, each controlled by the action of one enzyme. For instance, at least twenty enzymes are needed to break down the simple sugar glucose and make the energy available to the cell.

A classic example of an enzyme series is the one involved in the breakdown of phenylalanine and tyrosine, two of the twenty amino acids that are present in protein foods. Figure 9–3 shows some of the parts of this series and the consequences of an enzyme deficiency at several points in the series. **Phenylketonuria** or **PKU** results when there is a deficiency of the enzyme that converts phenylalanine into tyrosine. A liver enzyme normally makes this conversion, but in the absence of this enzyme the phenylalanine level of the blood goes up. PKU was discovered in 1934 by the Norwegian physician A. Fölling, who demonstrated that abnormal metabolites (phenylketones) were present in the urine of affected children. A fetus homozygous for the gene for PKU is normal before birth because the excess phenylalanine is removed by the placenta and converted in the liver of the mother. Within three days after birth, however, the phenylalanine level in the blood of these children will be measurably higher than in normal

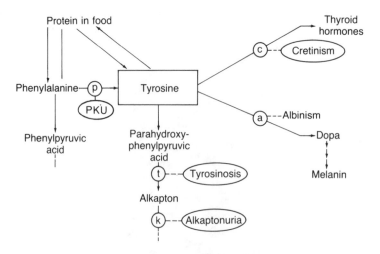

FIGURE 9–3. Series of enzyme-mediated reactions involving the amino acids phenylalanine and tyrosine. Inherited enzyme deficiencies have been found at five points in this series.

blood. Within a few weeks it will become as high as 60 or more milligrams per 100 milliliters of blood, as compared to a normal concentration of only about 2 to 3 milligrams. Also, there will be a rise in phenylpyruvic acid, a product produced from phenylalanine by another enzyme in an alternate pathway for its breakdown. The kidneys remove some of the excesses, but the levels remain high in the blood and interfere with brain development. It seems that, with so much phenylalanine in the blood, subnormal levels of the other essential amino acids are absorbed by the brain cells, which therefore cannot grow normally. Other body tissues may be affected for the same reason. The muscles and cartilages of the legs may be defective so that many individuals with untreated PKU cannot walk properly (see Figure 9–4). These people also are likely to have fairer than normal skin and hair because of a deficiency of tyrosine, which is needed for melanin formation.

Control of PKU is more difficult than control of galactosemia because phenylalanine is present in most protein foods and is one of the essential amino acids. We cannot eliminate proteins from the diet, but a synthetic mixture of amino acids is now available that includes all the essential amino acids except phenylalanine. If this is substituted for proteins soon after birth, an affected baby can develop relatively normally. Some phenylalanine is needed to build body proteins, but sufficient amounts can be obtained from other foods, such as fruits and vegetables, which are allowed along with the mixed amino acids. The object is to supply just enough phenylalanine for protein construction with little excess to accumulate in the blood. Damage can occur if a mistake is made either way.

Delayed recognition of PKU can result in brain damage before the diet can be started. The older test involved urine analysis, but the excess of phenylalanine could not be detected until about six weeks after birth, and by this time some brain damage had already begun. We now have the **Guthrie test,** which can make the detection as early as three days after birth. A drop of blood from the baby is placed on a disc of absorbent paper that is then placed on a culture medium seeded with *Bacillus subtilis,* a common bacterium. A chemical, thienylalanine, has been added to the medium and will inhibit growth of the bacteria unless phenylalanine is present to neutralize it. Normal blood has so little phenylalanine that there is only a very small area of growth around the disc. A heavy growth appears, however, around a disc impregnated with blood from a baby homozygous for the gene for PKU. Forty-three states now require this test for all babies. Even though only about 1 baby in each 10,000 born will have PKU, this one can be saved from a life of severe mental retardation if the condition is recognized early. Irrespective of the humanitarian aspects, mandatory testing can be a great financial saving to the state, which would otherwise have to care for the afflicted person in an institution.

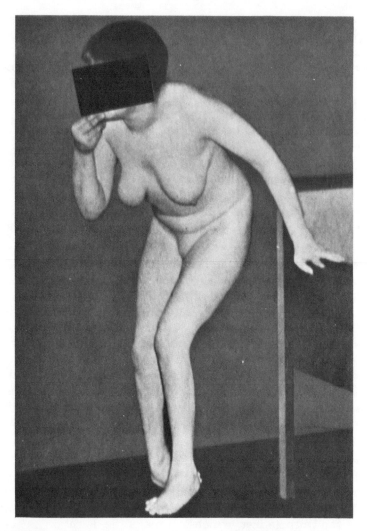

FIGURE 9-4. PKU in a woman who has inherited the inability to produce the liver enzyme needed for the breakdown of phenylalanine. Mental retardation and improper development of the leg muscles have resulted. (Courtesy Carl Larson.)

The special PKU diet can be discontinued at about nine years of age because by this time the brain is developed and cannot be damaged by high phenylalanine in the blood. A female, however, must be warned to go back on the diet should she ever become pregnant. Otherwise, the high level of this amino acid in her blood will damage the brain of her fetus. In fact, a few cases have been reported of brain damage to fetuses carried by heterozygous women. Some of these, especially those who have a large amount of protein in their diets, have a slightly elevated phenylalanine level of the blood, not enough to harm them as adults, but enough to affect the delicate brain tissue developing in a fetus. A test for the phenylalanine level of the blood may be added to the many other tests a woman has during early pregnancy. Women who show a somewhat higher than normal level may go on a low phenylalanine diet during pregnancy to avoid any possibility of damage to the brain of the fetus.

Tyrosine is another essential amino acid and may be obtained in protein food or from the conversion of phenylalanine. The way tyrosine is used in the cell depends on which enzymes act upon it. One enzyme leads to its use in the construction of structural proteins that are needed for growth and repair of protoplasm. Another uses it in the construction of the hormone, **thyroxine,** by the thyroid gland. The molecules of tyrosine are altered and combined with iodine to make this vital hormone that regulates metabolism. A recessive mutant allele of the gene that produces the enzyme needed for this synthesis fails to produce the enzyme in a functional form. A baby homozygous for this allele is normal at birth, thanks to the hormone produced by the mother, which can diffuse through the placenta into the fetus. Within a few weeks after birth, however, mental dullness and retarded growth become apparent. The thyroid gland may enlarge, and the child is said to have **genetic goitrous cretinism** (Figure 9–5). If no

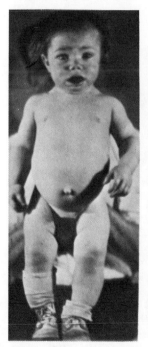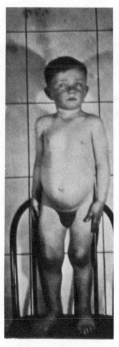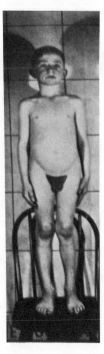

FIGURE 9–5. The child showing symptoms of cretinism, left, has been treated with thyroid extract and developed into a normal boy, right. (Courtesy Good Samaritan Clinic, Atlanta.)

treatment is given, the child will have greatly retarded mental, physical, and sexual development. This is another case, however, where genetic research has made it possible to overcome an inherited abnormality. Thyroxine, in the form of pills made from the dried glands of slaughterhouse animals, can provide the missing hormone and normal development will result.

Tyrosine may be acted upon by another series of enzymes in the skin, hair, and eyes, with **melanin** as the end product. **Albinism** results when an enzyme in this series is not produced (see Chapter 4). No way has been found to overcome this defect. Enzymes are proteins and protein molecules are too large to diffuse into the cells even if they are injected into the blood stream. Another type of albinism has been found that is due not to the deficiency of any of the enzymes, but to the inability of the cells to absorb tyrosine properly (Figure 9–6). With this deficiency, little melanin can be produced even though the enzymes are present. Albinos of this type have some melanin, but not enough for normal pigmentation. The type of albinism can be determined by culturing hair follicle cells in a solution rich in tyrosine. The enzyme-deficient type will still not produce any melanin, but the low-absorption type will show some pigment in the hair follicles.

Most of the tyrosine is acted upon by another enzyme that converts it into parahydroxyphenylpyruvic acid. Another enzyme in the series

FIGURE 9–6. An albino Navaho Indian girl at left, with her brother who has the pigmentation characteristic of members of the tribe. The girl's albinism is due to reduced tyrosine absorption by the skin cells. She can produce some melanin, but much less than normal. She is also sensitive to bright light because of the low melanin in the choroid coat of the eyes. The gene involved is very common among American Indians in the southwestern United States.

changes this into homogentisic acid (alkapton). Those individuals homozygous for a recessive allele of the gene that produces this latter enzyme have **tyrosinosis,** an apparently harmless condition that results in an excess of tyrosine in the blood and urine.

Another recessive gene fails to produce the enzyme that breaks down homogentisic acid. The result is **alkaptonuria,** which has already been described.

Causes of Enzyme Deficiency Diseases

Sometimes the substance normally acted upon by an enzyme will accumulate in high concentrations when the enzyme is missing. Galactosemia, PKU, and alkaptonuria are examples of the types of damage that can result. In other cases the body may suffer from the lack of a product produced by the enzyme (Figure 9–7). Cretinism and albinism are examples. Some of the symptoms of PKU also fall in this category. While some of the tyrosine used by the body comes directly from protein foods, much comes from the breakdown of phenylalanine. Denied this source, improper development of muscles and cartilages may result, as well as reduced melanin deposits and sometimes a reduced

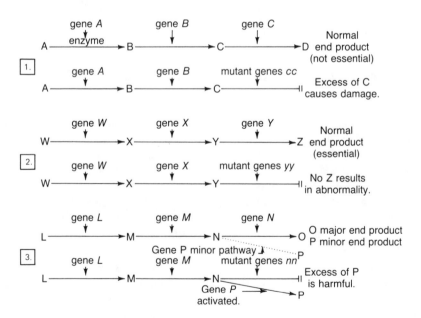

FIGURE 9–7. Enzyme deficiencies can cause abnormalities in three different ways. (1) Too much of an intermediate product that is not broken down. (2) Lack of an essential end product. (3) Too much of a product produced in an alternate pathway that is usually a minor pathway.

thyroxine output. In still other cases an enzyme deficiency results in an excess of a product of an alternate pathway. In a person with normal enzymes, only a little of the phenylalanine goes to produce phenylpyruvic acid, but when denied its major outlet to tyrosine, much of it goes into the usually minor pathway to phenylpyruvic acid. As a result, this acid accumulates in excessive quantities in the blood and may be partly responsible for the symptoms of PKU.

An upset in a feedback system is another way in which enzyme deficiencies can cause defects. One of the hormones secreted by the cortex of the adrenal glands is **cortisol.** A hormone from the pituitary gland in the brain, **ACTH (adrenocorticotrophic hormone),** stimulates the adrenal gland to produce precursors of cortisol, which are converted into cortisol by enzymes. The cortisol produced inhibits ACTH secretion. In this way the concentration of the two hormones remains in balance. A certain recessive gene, when homozygous, prevents formation of an enzyme necessary for the production of cortisol. As a result, the pituitary continues to secrete ACTH without inhibition and this stimulates excessive buildup of precursors of cortisol. These precursors break down, through an alternate pathway, and form substances with an androgenic (male-hormonelike) effect and cause what is known as the **adrenogenital syndrome.** In the female this takes the form of a masculinization of the genital organs and in the male there is a precocious sexual maturity.

Methods of Control of Enzymatic Defects

Now that scientists have determined the causes of defects resulting from enzyme deficiencies, they have been able to find ways of preventing these defects from developing. When the cause is an excess of a product not acted upon, withdrawal from the diet of that product or precursors of that product can sometimes prevent the defects. This can be done with galactosemia and PKU. In some cases, however, this is not possible. For instance, **Gaucher's disease** results from a deficiency of an enzyme for proper processing of lipids, which are a product of fat digestion. The result is an accumulation of fatty material in the spleen and liver, with death in the teens as these organs swell to enormous proportions. We cannot eliminate fats from the diet to correct this because fats can be manufactured in the body from other foods.

When a disease is caused by the lack of an end product, we can sometimes alleviate symptoms by administering the substance that the genes are not producing. Thyroid pills will prevent cretinism in those lacking the enzyme for thyroxine production. The pills contain a soluble hormone of low molecular weight that can be absorbed through the intestinal wall and by the cells as it circulates in the blood. People homozygous for the gene for **pituitary dwarfism,** however, cannot be stimulated into normal growth by taking the hormone from the pituitary gland. This hormone is of a high molecular weight and is too large to be diffused through cell membranes. For the same reason, we cannot correct albinism by administering melanin. Also, at the present time, there is little we can do to correct the diseases brought on by alterations of the feedback system.

To readers unfamiliar with cell physiology, it might seem that an obvious way to overcome all enzyme deficiency diseases would be to inject the missing enzyme. Unfortunately, however, enzymes are proteins, and proteins are such large molecules that they do not diffuse through the membranes around cells. We have learned, however, that cells may engulf larger particles, and in this way enzymes could enter cells. Encouraging results were obtained when enzymes were added to tissue cultured cells of children with **Tay-Sachs disease.**

Persons affected with Tay-Sachs disease cannot produce the enzyme **hexosaminidase A,** which is necessary for the proper metabolism of a class of lipids in the nervous system known as **gangliosides.** As a result, there is a gradual accumulation of these fatty gangliosides around cells of the brain. A baby with this condition is normal at birth, but by several months of age the accumulation begins to interfere with brain function and there is a progressive loss of mental and motor functions, with death by the age of three or four years. Hexosaminidase A is present in leukocytes of people carrying at least one of the normal dominant alleles. When the enzyme was added to tissue culture cells from children with Tay-Sachs disease, some of the enzyme was taken up by the cells, enabling them to process the gangliosides normally. These results gave hope that enzyme injection might be used on afflicted children to enable them to process the gangliosides in the brain cells. Unfortunately, however, this procedure failed to bring any improvement in the condition of children with Tay-Sachs disease. The injected enzyme seems to be rather quickly eliminated from the blood and any that may be engulfed by cells is broken down by other cellular enzymes before it can exert a noticeable influence.

In effect, at the present time the only control for Tay-Sachs disease is prenatal detection and abortion of affected fetuses. Since Tay-Sachs disease occurs primarily in Jews of Eastern European ancestry, and since heterozygous carriers of the Tay-Sachs gene can be identified in genetic screening programs, couples at risk for having affected children can be identified. In the event of a pregnancy, such couples could elect to have amniocentesis, and should the fetus be found to have Tay-Sachs, the couple then has the option of terminating the pregnancy.

Future Prospects for Overcoming Enzyme Deficiencies

Recent accomplishments in the methods of gene transfer offer exciting possibilities for treatment of those who are deficient in genes for the production of vital enzymes.

Transformation of Cells Through Chromosome Transfer

For nearly forty years we have known that bacteria can take up DNA from their surroundings and incorporate it into their own genetic systems, a process known as **transformation.** A strain of bacteria may lack a gene for the production of an enzyme necessary for the synthesis of a certain amino acid. DNA can be extracted from other bacteria that have this enzyme and placed on the culture medium upon which the deficient strain is growing. Part of this DNA will become incorporated into some of the bacteria and the bacteria will acquire the ability to synthesize the amino acid. More recently mammalian cells have been transformed in this manner. The Chinese hamster has a gene that codes an enzyme that can synthesize guanine, a base needed for DNA production. Because certain strains of mice do not have this gene, guanine must be included in their diet if they are to live. Chromosomes can be removed from hamster cells and placed in a medium on which tissue culture cells of the mice are growing. Some of the mouse cells will soon be able to survive on a medium without guanine. They have acquired the necessary gene from the hamster chromosomes.

Chromosomes from human cells have also been incorporated into mouse cells. The cells seem to engulf the chromosomes and, while much of an engulfed chromosome will be destroyed by cytoplasmic enzymes, sometimes a part of the DNA seems to reach the nucleus and become incorporated into the chromosomes there. Human tissue cultures might be started from a person with an enzyme deficiency disease and cells in this culture might be transformed with genes for the missing enzyme in chromosomes from a normal person, or from other mammals. Then the problem would be to reintroduce the cells into the donor so that the enzyme could be synthesized. It might also be possible to introduce chromosomes containing the needed gene directly into an organ where the enzyme is normally produced. For instance, chromosomes containing the gene for the enzyme that breaks down phenylalanine could be injected into the liver of a person with PKU. It is in the liver that this breakdown takes place and if some cells of this organ could be transformed into enzyme producers, the defect might be overcome.

Recombinant DNA Techniques

Dramatic developments in recombinant DNA or gene splicing research have taken place in recent years. Anyone who picks up a current *Time* magazine or a *Newsweek* is likely to find an account of the latest developments in this field. Some of these developments and the potential for using recombinant DNA techniques in correcting human genetic defects are discussed further in Chapter 12.

Gene Transfer by Viruses (Transduction)

Genes can be transferred from one bacterial cell to another by virus infections. A virus consists of a protein coat surrounding a core of nucleic acid, either DNA or RNA. When a virus infects a cell, the nucleic acid is injected and replicates within the cell. Then new protein coats are formed around the nucleic acid particles. Some of the DNA of the cell may also be included within the protein coats, and when the virus breaks from the cell and infects another cell, it transfers this DNA. Such transfer is known as genetic **transduction.** If all cells were destroyed by the infection, this principle would have no significance; but some viruses do not destroy all the cells they infect. Working with cultured cells taken from a person having galactosemia, scientists have used the technique of transduction to insert the gene for the missing enzyme, GPT. Some bacteria have the gene for galactose phosphate uridyl transferase (GPT), and when viruses from these bacteria infect cultured cells from persons with galactosemia, the cells acquire the ability to produce the enzyme. The cells are not harmed by the infection because this virus will not replicate in human cells. If we could then graft some of these cells back into the donor, it is possible that galactosemia could be corrected. Or, since the virus does not harm human cells, we might be able to inject it directly into the liver with the hope that some of the cells might incorporate the gene for the missing enzyme into their chromosomes.

Admittedly, these are all prospects that require considerable refinement of techniques, but considering the almost unbelievable advances in our ability to manipulate genes during the past twenty years, we can hope that some will be realized in the not too distant future.

PROBLEMS AND QUESTIONS

1   What is meant by an "inborn error of metabolism"?

2   Today we usually speak of "one-gene-one-polypeptide," but thirty years ago geneticists often referred to the "one-gene-one-enzyme hypothesis" of gene action. Cite two or three examples discussed in

this chapter where single genes do in fact control single enzymes. Can you suggest reasons why the one-gene-one-enzyme hypothesis has given way to the contemporary one-gene-one-polypeptide hypothesis?

3   Failure of fetal cells (obtained via amniocentesis) to produce the enzyme GPT indicates that the fetus has galactosemia; failure of such cells to produce hexosaminidase A indicates Tay-Sachs disease. Genetic counselors have reported that couples expecting a galactosemic baby will often allow the pregnancy to go full term and the baby to be born, while couples whose baby will have Tay-Sachs disease seem more prone to terminate the pregnancy. Speculate on the reasons why couples might react in different ways to having babies with these two different diseases.

4   In 1962 a baby girl was born who was screened for, and found to have, PKU. She was immediately placed on a low phenylalanine diet which allowed her to develop quite normally. During her junior year in college she married a classmate in whose family there was no history of PKU. Soon thereafter she became pregnant and subsequently gave birth to a severely brain-damaged baby. Various examining physicians suggested that the child was, no doubt, genetically "normal" with respect to PKU, but had all of the symptoms of PKU nonetheless. Explain. What are the probable genotypes of the parents and the child? How might this tragedy have been prevented?

5   In Washington, D.C., where the population is largely black, mandatory newborn screening for PKU has been discontinued because so few cases of the disease were identified. Suggest why the incidence of PKU may be so low in Washington as compared to the United States in general, where the frequency is variously reported as 1 in 10,000 to 1 in 15,000.

6   Both PKU and galactosemia can be controlled by dietary regulation. Can you suggest why the dietary regulation of PKU is more difficult than that of galactosemia?

REFERENCES

DAVIS, B. D. 1970. Prospects for genetic intervention in man. *Science* 170:1279–83.

FÖLLING, A. 1934. The excretion of phenylpyruvic acid (ppa) in the urine, an anomaly of metabolism in connection with imbecility. (Translation from the original German appears in *Papers on Human Genetics,* edited by S. H. Boyer. 1963. Prentice-Hall, Inc., Englewood Cliffs, NJ.)

GARROD, A. E. 1902. The incidence of alcaptonuria: a study in chemical individuality. *The Lancet* II:1616–20.

————. 1909. *Inborn Errors of Metabolism.* (Reprinted with a supplement by H. Harris, 1963.) Oxford University Press. London.

GILBERT, W., and L. VILLA-KOMAROFF. 1980. Useful proteins from recombinant bacteria. *Scientific American* 242 (4):74–94.

HARRIS, H. 1980. *The Principles of Human Biochemical Genetics,* 3d ed. North-Holland Publishing Co. Amsterdam.

SCRIVER, C. R., and C. L. CLOW. 1980. Phenylketonuria: epitome of human biochemical genetics. *The New England Journal of Medicine* 303 (23):1336–42.

Genes and Mutations

GENES ARE VERY STABLE; THEY CAN UNDER-
go countless thousands of replications with perfect copies resulting each time. On rare occasions,
however, a copy may be produced that varies
slightly from its progenitor. The situation can be
compared to a single mistake made by a typist
copying a sentence over and over again. One letter
may be changed, altering the meaning of the entire sentence. Likewise, a mistake in the replication of one base in the long DNA molecule of a gene
can have far-reaching consequences. Such a mistake is called a **mutation.** For example, a small
alteration in a gene that produces one of the blood
clotting factors occurred in Queen Victoria of England or one of her immediate ancestors. This
mutation caused her to have a son with hemophilia
(see Chapter 6). Now that we understand the chemical nature of genes and how their genetic code is
ultimately translated into protein products in the
cells, we can see how such mistakes in replication
can have drastic consequences. In this chapter we
shall review some of the information on the nature
of the genetic code and how alteration of this code
can take place. Also, we shall consider the very
important topic of agents that can increase the
mutation rate.

The Genetic Code

Espionage agents are proud of their skills in
cracking codes used by foreign governments to
send secret messages. During World War II agents
of the United States were able to crack the Japanese code and thus know about their plans for
warship distribution. Analogously, genes have a
code to convey information to the parts of the cells
that produce proteins. It has been a difficult code
to decipher because it is on the molecular level, yet

we have done it and now we know the genetic code for each amino acid.

Triplet Codons

Three bases in the DNA coding strand of a gene code for one amino acid. As the gene produces m-RNA, these three are transcribed into the three complementary bases. Such triplets of bases are known as **codons.** Sixty-four different codons can be constructed from the four kinds of bases (4^3 = 64). Since there are only twenty kinds of amino acids, this is more than enough to code for all of them. In fact, several different base combinations (codons) may code the same amino acid. Isoleucine, for instance, is coded by the m-RNA codons, AUU, AUC, and AUA. Table 10−1 shows the codons that specify each amino acid. Note that some amino acids are specified by as many as six triplet codons.

Comparison to Letters and Words

The genetic code for a polypeptide chain can be compared to a sentence. Each base is the equivalent of a letter and each codon is a word, with all words made of three letters each. Only four letters are used, but the

TABLE 10−1 The genetic alphabet

Codons in m-RNA	Amino Acid Coded	Codons in m-RNA	Amino Acid Coded
GCU GCC GCA GCG	alanine	AAA AAG	lysine
CGU CGA CGG CGC AGA AGG	arginine	AUG	methionine and initiator
AAU AAC	asparagine	UUU UUC	phenylalanine
GAU GAC	aspartic acid	CCU CCC CCA CGG	proline
UGU UGC	cysteine	UCU UCC UCA UCG AGU AGC	serine
CAA CAG	glutamine	ACU ACC ACA ACG	threonine
GAA GAG	glutamic acid	UGG	tryptophan
GGU GGC GGA GGG	glycine	UAU UAC	tyrosine
CAU CAC	histidine	GUU GUC GUA GUG	valine
AUU AUC AUA	isoleucine	UAA UAG UGA	terminator codons
CUU CUC CUA CUG UUA UUG	leucine		

"sentences" are very long and a different base substituted at one point in the "sentence" can change the meaning of a codon, thereby resulting in the substitution of the "wrong" amino acid at one point in the polypeptide chain. A polypeptide containing the "wrong" amino acid may be incapable of functioning normally. A written sentence in the English language must have an indication of the place to start reading and the place to end. A capital letter indicates the beginning point and several punctuation marks can indicate an ending. Likewise, the genetic code has an initiator codon (AUG), which also codes methionine. It signals the starting point only when it follows a certain leader segment of m-RNA. Three terminator codons (UAA, UAG, UGA) signal the place to stop production of a polypeptide chain. These three are known as nonsense codons; they code no amino acids.

Hemoglobin

Human hemoglobin was one of the first proteins for which the genetic code was deciphered, and serves as a good example of the method of protein synthesis.

Structure of the Hemoglobin Molecule

The hemoglobin molecule has a central core of **heme,** an iron-containing compound with four polypeptide chains, the **globulin,** radiating out from it (Figure 10–1). Normal **hemoglobin A** has two identical **alpha chains** of 141 amino acids each and two identical **beta chains** of 146 amino acids each. The m-RNAs would have an equal number of triplet codons, plus a terminator codon at the end of each. This would mean 142 triplets in the alpha-coding m-RNA, a total of 426 bases.

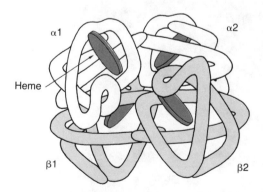

FIGURE 10–1. The hemoglobin molecule consists of two identical alpha chains each consisting of 141 amino acid residues and two identical beta chains each consisting of 146 amino acid units. Each chain has associated with it an iron-containing heme group (disc-shaped object).

Beta-coding m-RNA would have 147 triplets, 441 bases. By electrophoresis and chromatography, the sequence of amino acids for each of these polypeptide chains has been determined and, since we know the codons for each amino acid, we can determine the sequence of triplets in the m-RNA and the genes that code these chains.

Variant Forms of Hemoglobin

Alpha chains are found in the hemoglobin at all ages, but not beta chains. Blood of a very young embryo has a pair of **epsilon chains** instead of beta chains. These, along with the alpha chains, form what is called embryonic hemoglobin, **hemoglobin E,** which is produced in the yolk sac. The spleen and liver are the next to take over hemoglobin production, and **gamma chains** are substituted for the epsilon chains. This is fetal hemoglobin, **hemoglobin F.** As the bones are formed, the red bone marrow takes over the production of red blood cells and hemoglobin A is found in these cells. The times when these three kinds of hemoglobin are produced are shown in Figure 10–2. In addition, note that there is a fourth type of hemoglobin, which has two **delta chains.** This is **hemoglobin A_2.** Only about 2.5 percent of adult hemoglobin is hemoglobin A_2. Thus, it is apparent that four different genes are involved in producing the second pair of polypeptide chains. Each of these four is strongly expressed at one time of life and then becomes dormant when another gene takes over.

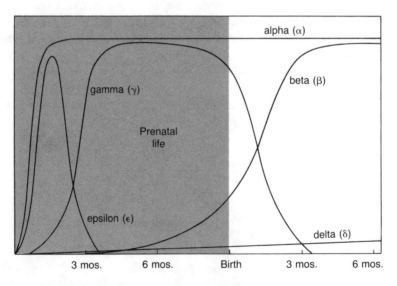

FIGURE 10–2. Different types of hemoglobin and the time of their production. The alpha chain is produced at all ages, but the second pair of chains varies according to the time of life.

Mutations of Genes for Hemoglobin

One mutation of the gene Hb^A, which produces the beta chain of hemoglobin A, gives the allele Hb^S, which produces **sickle-cell hemoglobin** (see Chapter 4). Figure 10–3 shows how these two kinds of hemoglobin can be recognized by electrophoresis. **Hemoglobin S** moves more slowly toward the positive pole than hemoglobin A does. It has a difference in one amino acid in the beta chain (valine has been substituted for glutamic acid) at the sixth position in the 146 amino acid chain. This difference alters the electric charge of the entire molecule because glutamic acid has an overall negative charge, while valine is neutral. Since the two alleles are codominant, the heterozygous person produces both hemoglobin A and hemoglobin S.

Another mutant form of the Hb^A was found in West Africa. This mutant allele, Hb^C, causes a substitution of lysine for glutamic acid at the same point on the beta chain and produces **hemoglobin C.** Since lysine is positive, hemoglobin C migrates toward the positive pole even more slowly than does hemoglobin S.

With the preceding information, we can speculate on the method of mutation that has resulted in these different alleles. One of the m-RNA codons for glutamic acid is GAA, which corresponds to CTT on the coding strand of the gene. If one base pair were changed in the gene (see Table 10–2) so that the side of the gene that codes m-RNA became CAT, the codon produced on the m-RNA would be GUA, which is a codon for valine. Thus, the gene Hb^A would be changed to its allele, Hb^S. Selection would favor this mutant allele in areas where malaria was prevalent. The mutant allele Hb^C could have arisen by a substitu-

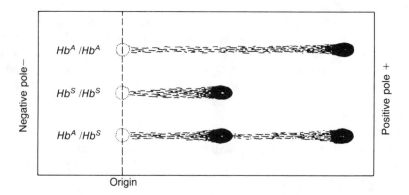

FIGURE 10–3. Electrophoresis brings out the differences in the two kinds of hemoglobin. When a drop of blood is placed on paper and exposed to an electrical field, hemoglobin S moves more slowly toward the positive pole than does hemoglobin A.

tion of T for C in the same triplet of the gene. The sequence would then be TTT that would produce the m-RNA codon AAA that would substitute lysine at this point on the beta chain. Hemoglobin C also gives some resistance to malaria.

TABLE 10–2   Three variants of adult hemoglobin*

Hemoglobin Types	Amino Acid in Position 6 of Beta Chain	DNA Triplet	m-RNA Codon
S	valine (neutral)	CAT	GUA
		↑	
A (normal)	glutamic acid (negative)	CTT	GAA
		↓	
C	lysine (positive)	TTT	AAA

*Arrows indicate how a change of one base of one codon in the gene for the beta chain can alter the entire hemoglobin molecule.

In addition to inversion and substitution of base pairs, other alterations of a gene can result in mutations. One base pair may be deleted or transferred to another position. Some changes might result in a terminator sequence in the middle of the code and the result would be a shortened code on the m-RNA that would produce polypeptide chains that would lack all the amino acids from that point on.

As routine electrophoretic analyses of hemoglobin from many people were done, many other mutant variations in the polypeptide chains were found. In fact, we have now discovered amino acid substitutions at all but a few positions of alpha and beta chains. Most of these do not alter the function of hemoglobin or the appearance of the red blood cells. This indicates that many mutations must occur that change the amino acid sequences in proteins without causing substantial phenotypic change. These alterations are in the parts of the chains that are not vital to the functioning of the protein. In addition, there must be many mutations that do not alter the amino acid sequence. For instance, AAA on the gene codes phenylalanine, and if a substitution changes this to AAG, it would still code phenylalanine because AAG is an alternate codon for this amino acid. Such mutations cannot be detected by present techniques.

Kinds of Mutations

Mutations can be classified according to the nature of their expression in the following ways.

Visible Mutations

Visible mutations produce an easily seen or demonstrated phenotype. They are better known than any of the others, but actually represent no more than about one percent of the total number of mutations that occur. If visible mutations are dominant, codominant, or intermediate they can be detected in the first generation after they appear. Recessive mutations are much more common, but it may be many generations before a recessive mutant is combined with another like itself and is expressed. X-linked recessives are an exception, of course, for they can be expressed in the first generation sons when they appear in a woman and can appear in grandsons of a man in which they occur.

Lethal Mutations

Many mutations cause such extreme alteration of body processes that they are lethal when expressed. These are more frequent than visibles. A dominant lethal mutation will, therefore, be eliminated in the first generation after its appearance, with the exception of the few that cause death after sexual maturity. Most lethals are recessive and can be spread widely in a population after they arise as a mutation. It is estimated that each person carries an average of two or three such lethal genes.

Detrimental Mutations

Many mutations have an effect that is too small to be easily recognized, but which can cause a slight reduction in vitality and ability to survive. The liver may not remove toxic substances as efficiently, or the kidneys may not respond readily to the concentration of dissolved substances in the blood. While any one detrimental mutation will have only a small effect, the combined expression of many such genes can add up to a major reduction in vitality. The offspring of inbred domestic animals tend to be less vigorous and to have a shorter life span than do those that come from outbreeding. The number of recessive detrimentals that become homozygous is much greater in the offspring of closely related parents.

Neutral Mutations

As has already been described, electrophoresis has shown that many mutations occur that have no noticeable phenotypic effect. Enzymes, for instance, have an active site that accomplishes the function of the enzyme. If the active site is not modified by the mutation, the enzyme may continue to function normally. A mutation that alters an amino acid in the enzyme can make the enzyme nonfunctional and have a marked phenotypic effect, if the substituted amino acid results in a change in the shape of the enzyme's active site.

Somatic Mutations

Thus far we have concentrated our attention on mutations that arise in reproductive cells. These are the only ones that can be transmitted to offspring, but the genes in somatic or body cells can mutate also. Most somatic mutations are never detected. You may be heterozygous for albinism and will have normal pigmentation because the gene for melanin production is dominant. If, while you are reading these lines, the dominant allele mutates to the allele for albinism in a cell in your intestine, this cell becomes homozygous. Even though this cell divides and forms a clone of homozygous cells, there is no effect. Melanin is not produced in cells of the intestine. If such a mutation took place in a cell of your skin, you might get a tiny albino spot after the cells multiplied for a time, but you might never notice it.

Somatic mutations that occur in cells of the early embryo, however, are another matter. Each cell can multiply and produce a large mass of tissue that may easily be noticed in the baby. Figure 10–4 shows a young man with an island of albino tissue around one eye that could have come about as a result of a mutation in the early embryo.

FIGURE 10–4. A small island of albino tissue includes the upper eyelid, part of the eyebrow, and surrounding skin in this young man. A somatic mutation in one cell during early embryonic development could have been the cause of this mosaic effect.

Somatic mutations may be involved in cancer induction in some instances. We know that genes regulate the growth and differentiation of cells and a mutation of one of these might lead to unregulated growth and dedifferentiation. Support for this concept comes from the observation that the same agents that cause mutations in reproductive cells are also very likely to be carcinogenic (cancer causing). See Chapter 14 for further details on the relation of genetics to cancer induction.

Frequency of Mutation

How often do mutations occur? This can easily be determined for dominant mutations by a tabulation of the number of babies that express the mutation when neither parent had the trait. Hospital records in Copenhagen showed that 14 babies out of 128,763 born had **chondrodystrophic dwarfism.** The head and trunk is about normal size, but the arms and legs are greatly shortened (Figure 10–5). Three of the parents had this type of dwarfism, so 11 of the cases in the babies must have been the result of new mutations. This is a rate of 1 in each 11,500 births, or 1 mutation in each 23,000 parents. If we assume that the average age of the parents was thirty years, the frequency of this mutation would be about once in each 690,000 years. The frequency of X-linked recessive mutants also can be rather easily estimated because they appear in the sons of women in which the mutation has occurred. Pedigree studies can give some indication of the frequency of autosomal recessives. Table 10–3 shows the rates of mutation for some genes. Since these represent only a few of the visible mutations, which are many fewer in number than the other kinds, we can see that the chance of any reproductive cell carrying some sort of mutation is rather high.

Mutagenic Agents

Various natural agents are constantly producing mutations, but these occur at a very low frequency. Certain agents of human origin can bring about an increase in this frequency. These mutagenic agents may also result in an increase in the frequency of certain chromosome aberrations. In recent times the accelerated use of various chemical and physical agents that may have mutagenic properties has caused concern among some geneticists who fear possible significant damage to the genes and chromosomes that will be passed to future generations. Let us see what some of these agents are and how they might cause genetic damage.

FIGURE 10—5. Chondrodystrophic dwarfism, a dominant trait character-ized by shortened and bowed legs and arms, is expressed in this young man even though neither parent had the trait. Hence, the gene for dwarfism may have appeared in a gamete of one of his parents as a new mutation.

Ionizing Radiation

There are many different kinds of radiation; visible light rays, infrared rays, ultraviolet rays, and radio and TV signals are familiar examples. The kind that causes concern among geneticists, however, is ionizing radiation, that which causes ionization in matter through which it passes. Such ionization can cause changes in both genes and chromo-somes. **X rays** and radiation from **radioactive isotopes** are the most important types of ionization-inducing radiation.

In 1927 H. J. Muller found that X rays increase the rate of mutation as well as chromosomal aberration in fruit flies. Continued investiga-tion has shown that this effect applies to many forms of life, from bacteria to mice. Other forms of ionizing radiation have proved to be mutagenic as well. Since genes are constructed in the same basic fash-

TABLE 10—3 Spontaneous mutation rate

Traits	Mutations per Million Gametes per Generation
autosomal dominants	
chondrodystrophic dwarfism	42
epiloia (Butterfly rash on face; tends to become cancerous)	6
anirida (absence of iris of eyes)	5
Marfan syndrome	5
micropthalmia (eyes reduced in size and nonfunctional)	6
multiple polyposis (polyps in colon; often become cancerous)	13
nail-patella syndrome (defects of nails and bones)	2
retinoblastoma (tumors on retina of eyes)	23
autosomal recessives	
albinism	28
epidermolysis bullosa (blisters on feet)	50
X-linked recessives	
hemophilia A	32
muscular dystrophy, Duchenne type	43

ion in all forms of life, it appears certain that the genetic material of humans is also changed. In fact, we have some direct evidence that this is true. While there is a difference of opinion as to how much ionizing radiation poses a genetic hazard, practically all those with a knowledge of genetics agree that we should keep exposure to the minimum based on medical needs.

Medical use of ionizing radiation has expanded greatly in recent years. There has also been increased exposure to radiation from isotopes produced in nuclear generating plants and in tests of nuclear weapons. The possibility of genetic damage, however, has prompted attempts to restrict exposure. New techniques for making X-ray diagnosis have made it possible to reduce greatly the amount of radiation used at each diagnosis. Many former uses of X rays and isotopes in medicine have been discontinued where other methods of diagnosis and treatment are available. Mobile chest X-ray units and routine fetal X-ray examinations are examples. Improved handling of isotopes and wastes from nuclear generating plants has reduced radiation exposure from that source. International bans on atmospheric testing of nuclear weapons have greatly reduced the exposure to atomic fallout worldwide. (See Chapter 14 for further information on the genetic effects of ionizing radiation.)

Mutagenic Chemicals

Some chemicals have been found to be mutagenic. These include such widely divergent substances as mustard gas (the first chemical shown to be mutagenic), formaldehyde, ethyl urethan, acridine dyes, nitrous acid, epoxides, phenol, manganous chloride, bromouracil, dioxan, and even caffeine and theobromine, which are so widely consumed in coffee, tea, and chocolate. Many of the investigations have been made on bacteria, molds, and tissue cultures of cells of humans and other higher animals, so we cannot conclude that all of these chemicals are mutagenic in humans in the concentrations in which they are usually found. But they are suspect.

Drugs

The widespread use of drugs for their hallucinatory effects has caused concern, not only because of the physiological effects on the individual users, but also because of possible effects on future generations. Lysergic acid diethylamide (LSD) has been shown to increase the number of chromosome aberrations in human tissue cultures and some reports indicate an increase in the number of aberrations in blood leukocytes as well. One study showed an increased rate of mutation when LSD was added to the food media of the vinegar fly, *Drosophila*. It also seems to be teratogenic, causing birth defects when used extensively by pregnant women, and so is also suspected of being mutagenic. Marihuana has been found to inhibit the immune response of leukocytes in the blood of chronic users; agents that are immunosuppressants are often also mutagenic, but no final conclusions have been established. Certain antibiotics in the streptomyces group have also come under suspicion.

Temperature

In general, as the temperature rises, mutation rates also rise. This has been clearly demonstrated in experimental organisms such as the vinegar fly, *Drosophila*. The effect of the increase in temperature on mutation rate is not surprising since the rate of chemical reactions is generally increased by a rise in temperature. In all likelihood, temperature affects both the rate of reaction of DNA with other chemicals in the cell, as well as the stability of the DNA molecule itself. One study of humans suggests that the wearing of clothing by human males increases scrotal temperature and could thereby cause an increase in mutation frequency.

Detection of Induced Mutations

With this great array of potential mutagenic agents to which we might be exposed, how are we to know which pose a true genetic hazard and are to be avoided? Some of the investigations have been made with concentrations of the agents much higher than those to which people are normally exposed. Also, many studies were made on tissue cultures where the contact with the agents was much longer than in a living system where the natural body processes eliminate many of the agents rather quickly. Hence, we need results from studies of the genetic effects on humans or, at least, on closely related forms of life. Such studies are very difficult, however, because the natural mutation rate is so low that many people or other animals must be studied in order to try to detect any increase in those exposed to the agents. There are, however, ways we can gain clues to the possible damage to humans.

A decrease in the male to female ratio of children born to women exposed to a suspected mutagenic agent might indicate an increase in X-linked lethal mutations (see Chapter 6 for results of some such studies). Differences in the birth weight of babies born to exposed and unexposed parents could indicate an increase in detrimental mutations. Infant mortality and viability are other indicators of possible genetic damage. An increase in the number of spontaneous abortions and stillbirths from exposed parents suggests induced lethal mutations. In addition, the technique of determining the amino acid sequence of polypeptide chains that make up body proteins can reveal an increase in the number of neutral mutations, which seem to be the most common of all.

Agents are suspected of being mutagenic if they have teratogenic effects that are comparatively easy to confirm (see Chapter 3). Ionizing radiation in early pregnancy is teratogenic, so most physicians do not make X-ray photographs during pregnancy unless it is a medical necessity. Many will give radiation treatments to mature women only during the first two weeks after menstruation because that is the one time they can be almost certain that there is no pregnancy. Many pregnant women who were exposed to radiation from the atom bombs in Japan gave birth to defective babies and many others had abortions, because the fetus is much more sensitive to radiation damage than is an adult. Many pregnant women in Vietnam had defective babies after they had been exposed to herbicides used to defoliate the jungles. These various agents that have been shown to be teratogenic are either known or suspected of being mutagenic. Carcinogenic agents and immunosuppressants are also suspected of being mutagenic (see Chapter 14).

To sum up, people are now being exposed to more and more new substances. Some of these agents have proved to be mutagenic, and others are suspected. We do not yet have full information on the impact of these agents on future generations, but since changes in genes and chromosomes cannot be undone, we should be cautious about any unnecessary exposure to those for which there is strong evidence of a mutagenic effect. Any errors in judgment should be on the side of caution.

PROBLEMS AND QUESTIONS

1 **(a)** DNA $\xrightarrow{\text{(b)}}$ RNA $\xrightarrow{\text{(c)}}$ Polypeptide $\longrightarrow$ Trait
   Using the terms "replication," "translation," and "transcription" as they were used in this chapter, associate the appropriate term with each of the lettered steps in the sequence shown above.

2 "Normal" adult hemoglobin (hemoglobin A) has glutamic acid as the sixth amino acid in the β polypeptide chain. Hemoglobin S, sickle cell hemoglobin, has valine instead of glutamic acid in the sixth position. CTT and CTC are DNA triplets that code for glutamic acid. Using the genetic alphabet in Table 10–1, determine what single nucleotide substitutions must occur in each of these triplets so that they will specify valine instead of glutamic acid.

3 Suppose that . . . CAT TAC GAT GAG . . . represents a portion of the DNA coding strand for a particular gene. Note that the segment given consists of four codons. Now suppose that the DNA is exposed to a chemical that causes the addition of a nucleotide containing base G after the first DNA triplet so that the sequence now reads . . . CAT GTA CGA TGA G＿＿ . . . . What effect would such a "frameshift" mutation have on the polypeptide product produced by this gene?

4 Would you expect a frameshift mutation such as described in Question 3 above to be more or less serious in its consequences than a mutation such as the one that causes sickle hemoglobin? Justify your answer.

5 Using the genetic alphabet in Table 10–1, explain why the mutation shown below might be called a "silent" mutation.

   DNA coding strand: TAC GGG TTC CTT CGA
   $\downarrow$
   mutation: A
   Mutated DNA strand: TAC GGA TTC CTT CGA

6 Many pollutants produced by industry have been the cause of considerable concern to environmentalists. Individuals and groups concerned about environmental quality tend to emphasize that the pollutants are toxic to humans and to plant and animal life. They may also mention that some of the pollutants are carcinogenic or that they are teratogenic, producing developmental abnormalities in the newborn. Why do geneticists also have cause to be concerned about these polluting chemicals?

7 The text suggests that a dominant lethal mutation may be expected to be eliminated in the first generation after its appearance by the death of individuals in whom it occurs. Huntington's disease is regarded as a dominant lethal, yet it survives in the population and is not immediately eliminated. Explain this apparent contradiction.

REFERENCES

DISHOTSKY, N. I., W. D. LOUGHMAN, R. E. MOGAR, and W. R. LIPSCOMB. 1971. LSD and genetic damage. *Science* 172:431–40.

EHRENBERG, L., G. V. EHRENSTEIN, and A. HEDGRAN. 1957. Gonad temperature and spontaneous mutation-rate in man. *Nature* 180:1433–4.

GENETICS CONFERENCE, COMMITTEE ON ATOMIC CASUALITIES, NATIONAL RESEARCH COUNCIL. 1947. Genetic effects of the atomic bombs in Hiroshima and Nagasaki. *Science* 106:331–33.

MULLER, H. J. 1927. Artificial transmutation of the gene. *Science* 66:84–87.

PERUTZ, M. F. 1964. The hemoglobin molecule. *Scientific American* 211(5):64–76.

——. 1978. Hemoglobin structure and respiratory transport. *Scientific American* 239(6):92–125.

VOGEL, F., and G. RÖHRBORN, Editors. 1970. *Chemical mutagenesis in mammals and man.* Springer-Verlag, New York.

Chromosomal Aberrations

MITOSIS AND MEIOSIS ARE USUALLY VERY exact processes that result in an even distribution of chromosomes to the daughter cells. As in most biological processes, however, occasional mistakes are made; chromosome aberrations result from these mistakes. It is estimated that over 7 percent of all conceptions result in a zygote with some kind of chromosome abnormality. Most of these have a lethal effect early in development. Studies of cells from spontaneously aborted fetuses indicate that about 50 percent of them have chromosome aberrations. Some aberrations, however, are not so severe as to be lethal, but the babies born with them are usually greatly handicapped. Some studies suggest that 1 of 200 liveborn babies has a chromosome aberration; others estimate the frequency to be as high as 0.7 percent.

Chromosome aberrations fall into two categories: those involving the number of chromosomes and those involving the structure of chromosomes.

Aberrations Involving Alterations in Chromosome Number

During meiosis a pair of chromosomes may adhere so tightly that they do not pull apart at the anaphase, so both go to the same pole. As a result of this nondisjunction, duplicate chromosomes go to one daughter cell and none of this type of chromosome to the other. In Chapter 5 we learned how nondisjunction of sex chromosomes causes various abnormalities related to sex. When nondisjunction of autosomes occurs, serious abnormalities develop because of the alteration in the balance of genes in the cells.

175

Trisomy and Monosomy

When a gamete carrying two chromosomes of one kind unites with a normal gamete, the zygote will have three of one kind of chromosome, but two of each of the other kinds. This condition is known as **trisomy.** A baby with trisomy-18 has three chromosomes 18. Zygotes formed from a gamete lacking a chromosome would be **monosomic** and are designated as monosomy-18, etc. Trisomic and monosomic zygotes are said to be **aneuploid;** aneuploids differ from the normal diploid with respect to one, or perhaps two or three, chromosomes. All aneuploid zygotes give rise to embryos of low viability. While all the genes may be normal, the balance between the genes has been upset, which can have far-reaching consequences. With the exception of embryos monosomic for the X, almost all monosomics die very early in embryonic existence. Even so, most monosomic X embryos also abort; perhaps only 2 percent reach full term. Most trisomics also abort, but a few are born alive, although they have a low viability after birth. Trisomics for sex chromosomes fare much better, probably because the inactivation of all but one X in the Barr bodies maintains the normal chromosome balance. Three autosomal trisomies appear with predictable frequency in newborns.

Trisomy-21 (Down Syndrome, or Mongolism)

Characteristics of trisomy-21 include a skin fold of the upper eyelids, short stature, broad hands and stubby fingers and toes, a wide, rounded face, a large protruding tongue that makes coherent speech difficult, and mental retardation (see Figure 11–1). Respiratory infections, leukemia, and heart defects are also common among those with this syndrome. In the past, about half of these individuals died during their first year, and most of the remainder did not live beyond twenty years. Now, with open heart surgery, antibiotics, and other medical techniques, the life span has been greatly extended to an average age of about forty years.

Many theories to explain Down syndrome were proposed before a chromosome aberration was clearly identified as the cause. Langdon Down in the nineteenth century suggested that the syndrome was a throwback to some ancient Mongolian ancestors. Then, when it was noted that the chance of having this type of child increased with the age of the mother, the cause was assumed to be some sort of degenerative changes in the female reproductive tract associated with aging. Shortly after Tjio and Levan greatly improved the tehniques for identifying human chromosomes in 1956 (see Chapter 1), the true cause was discovered. Those with the syndrome typically had three of one of the very short chromosomes, later designated as 21 (Figure 11–2).

The average risk of having a child with trisomy-21 is about 1 in 750 to 800, but it is only about 1 in 1500 for women in their early twenties,

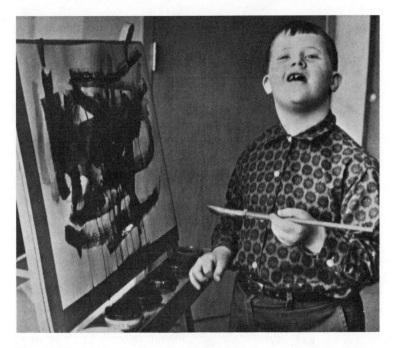

FIGURE 11–1. Down syndrome, or mongolism, is expressed in this boy because he has trisomy-21; that is, he has three chromosomes 21. He is receiving special training to enable him to achieve his maximum potential within the limits imposed by the extra chromosome.

climbs to about 1 in 70 for those in their early forties, and jumps to about 1 in 25 in those forty-five or older. About 25 percent of all babies with this syndrome are born to women over thirty-five but only about 5 percent of the total babies born have mothers who have passed this age. If all women would complete their families before age thirty-five the frequency of the syndrome would drop by about 25 percent. Many physicians recommend amniocentesis for all women who become pregnant after thirty-five. Cells from the amniotic fluid will show trisomy-21 if it is present, and many prospective parents will choose therapeutic abortion over the tragedy of having a child with this defect.

Why should the incidence of trisomy-21 increase for older mothers? Several factors may be involved. The chromosomes pair in prophase of oogenesis but do not separate until about the time of ovulation, which will be from about twelve to fifty years later (see Chapter 2). Chromosomes that remain synapsed a longer time might disjoin less readily than those with a short period of contact. Also, older women are more likely to have had greater exposure to ionizing radiation, viruses, and other agents that are known to increase the rate of nondisjunction. Delayed fertilization has also been suggested as a possible factor. Ex-

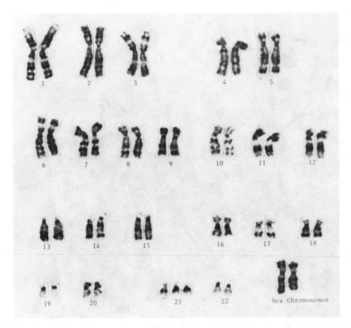

FIGURE 11–2. Karyotype of a person having trisomy-21 Down syndrome. Note that three chromosomes 21 are present in this female, resulting in 47, XX + 21. Chromosomes stained using the G-banding technique. (Courtesy Catherine G. Palmer, Indiana University School of Medicine.)

periments on other animals show that once an oocyte or egg is mature, its chance of producing an abnormal offspring increases with elapsed time before fertilization. This is probably because of an increase in nondisjunction as well as other chromosome aberrations at the second meiotic division, which does not occur until after fertilization. Older women, with their changing hormone balance, tend to have more delayed ovulations than do younger women. Also, older women are likely to have intercourse less frequently, as a rule, than younger women, so the chance of a delay between ovulation and fertilization increases. The egg has only a day or two after ovulation during which fertilization can occur, but even a delay of this short a time could be a factor. Still another plausible suggestion is that the chance of fetal survival seems to be greater in older women, especially if they have had previous pregnancies. The number of multiple births is greater in this group because there are fewer losses to abortion when more than one fetus occupies the uterus. Most trisomy-21 fetuses are aborted in all age groups, but perhaps more of them survive to birth in the older women. These various possible factors are not mutually exclusive and more than one might be involved.

What about the fathers? Can they not also be the source of the extra chromosome 21? They can, but not as frequently as the mothers. The fluorescent, Q-banding technique (described in Chapter 1) makes it possible to determine the extent of paternal involvement. Some chromosomes, including 21, vary in their regions of low DNA content (heterochromatin), and these glow brightly under the ultraviolet light microscope. It is, therefore, often possible to identify the source of the extra chromosome 21 (see Figure 11–3). When two of the three 21s show fluorescent patterns similar to those of the father, then it is evident that nondisjunction occurred in the father. One study of 18 children with Down syndrome and their parents showed that the extra chromosome came from the father in 8 cases, and from the mother in the remaining 10. These figures are too small to establish reliable proportions, but other studies show that paternal nondisjunction is involved in about 25 percent of the cases. Studies published in the late 1970s suggest that paternal age may also be related to the incidence of Down syndrome. Men over fifty-five years old show an increased likelihood of producing children with trisomy-21.

Can heredity play any part in the chance for Down syndrome? If parents already have a child with the syndrome, are they likely to have affected children in the future? This is one of the most common questions asked of genetic counselors. Nondisjunction seems to be a chance occurrence, so the answer might seem to be negative, but sad experience has shown that this is not always true. Statistical studies suggest

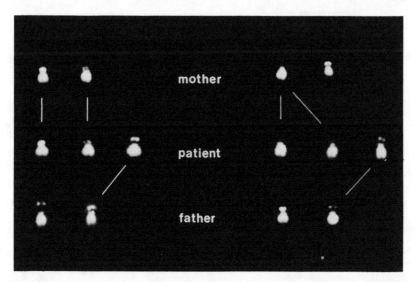

FIGURE 11–3. The fluorescent Q-banding technique can show from which parent a trisomy-21 child has received the extra chromosome. Chromosomes 21 exhibit morphological differences, especially in their satellites. In the case on the left it is evident that the two chromosomes 21 came from the mother as a result of nondisjunction in the first meiosis. In the case on the right the two 21s also came from the mother, but since they are alike, the nondisjunction must have been in the second meiosis.

that the chances are increased. For a young woman, the chance is about 1 in 1500 if none of her previous children have had the syndrome, but the chance increases to about 1 in 100 after a child with the syndrome has been born to her. Some factors other than chance must be operating and these factors could be more common in some women than in others. They could either be internal physiological factors or environmental agents that induce nondisjunction. In general, the recurrence risk appears to be about 1 in 100, regardless of the mother's age.

There is another way that Down syndrome can arise that is definitely influenced by heredity. In about 4 percent of Down syndrome cases the chromosome number is the normal 46 rather than 47. A karyotype, however, will show that a third chromosome 21 has been translocated onto one of the other chromosomes, often a chromosome 14 (see Figure 11–4). A person with this translocation and a single 21 will be normal, even though only 45 chromosomes are present. Such a person, however, has a high probability of bearing a child with Down syndrome (see Figure 11–5). Of the various possible zygotic combinations, about one-third of those that produce fetuses that can survive will be expected to have the syndrome. Actual tabulation, however, shows that

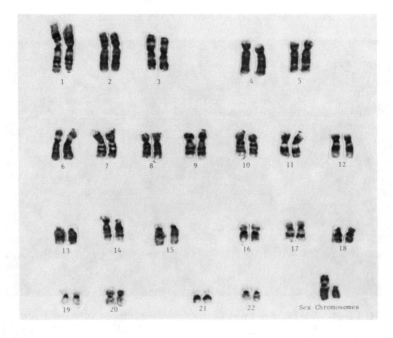

FIGURE 11–4. Karyotype of a male with Down syndrome resulting from a chromosome 21 translocated to a chromosome 14. The G-banding technique has been used. (Courtesy Catherine G. Palmer, Indiana University School of Medicine.)

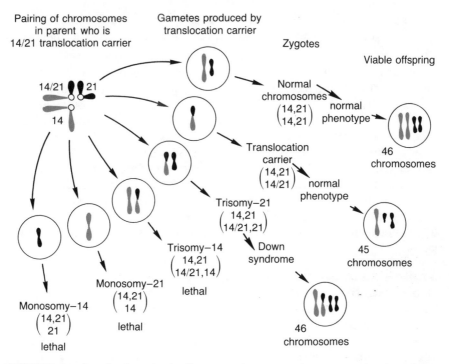

FIGURE 11–5. A phenotypically normal parent who carries a 14/21 translocation can produce six kinds of gametes relative to chromosomes 14 and 21. Assuming these gametes unite with a normal gamete containing one chromosome 14 and one chromosome 21, six kinds of zygotes can be produced. Only three of the zygotes will be viable.

only about 11 percent of the children born to women with this translocation have the syndrome. This reduction from the expected 33 percent might be accounted for, in part at least, by the higher rate of abortion of fetuses with the syndrome. Also, fewer gametes might reach maturity when they carry the translocation, a principle known as **meiotic drive.** It is known that abnormal chromosome complements slow the process of meiosis and gamete maturation. This seems to be of greater importance in spermatogenesis than in oogenesis because only about 2 percent of the children fathered by male translocation carriers have the syndrome. Maturation of spermatocytes and spermatids with the translocated chromosome must not proceed as well as maturation of oocytes in oogenesis.

A few cases have been found where the two 21s become attached to each other and never disjoin in meiosis. Each gamete receives either two 21s or none. All the babies born to a parent with such an attachment will have Down syndrome; fetuses that come from a gamete with no 21s are aborted early. This is a self-eliminating type of aberration since people with Down syndrome usually do not have children, so it would affect only the children of persons in which the aberration appeared.

Of all mothers having Down syndrome children, the proportion of translocation Down's is highest among young women. The chance for a translocation is the same for both, but the proportion is greater for young women because they have fewer cases due to nondisjunction. About 9 percent of Down's children born to women under thirty years of age have the translocation form of the syndrome, but only about 1.5 percent of Down's babies born to mothers over thirty years old have the translocation form. All parents who have had a child with Down syndrome should have a karyotype made of their chromosomes before considering another pregnancy to determine if a translocation is involved. This is especially important when the mother is in the younger child-bearing age group.

Trisomy-18 (Edwards Syndrome)

Babies with trisomy-18 are smaller than normal, have a feeble cry, a mottled skin, low-set and deformed ears (see Figures 11–6 and 11–7), a rounded bottom of the foot (rocker-bottom foot), an index finger that tends to overlap the third finger, heart and kidney abnormalities, skeletal deformities, and severe mental retardation. Their average survival time is only about ten weeks. Perhaps 1 child in each 5000 live births has this syndrome. Many fetuses that are trisomic-18 abort spontane-

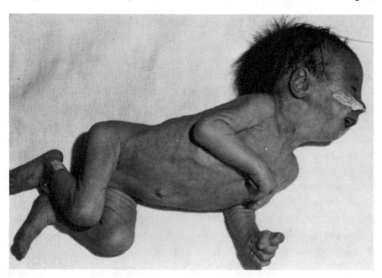

FIGURE 11–6. Newborn with trisomy-18, Edwards syndrome. Notice the peculiar manner in which the child clenches his fist; this trait is characteristic of the syndrome. (Courtesy Richard C. Juberg, M.D., Ph.D., The Children's Medical Center, Columbus, Ohio.)

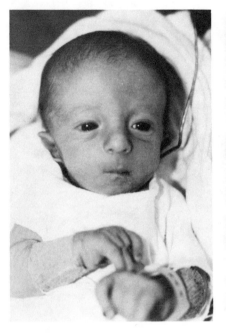
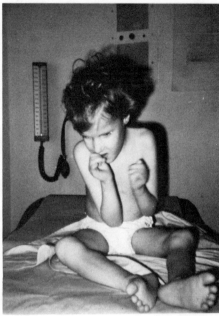

FIGURE 11—7. Newborn (left) and nine-year-old (right) with trisomy-18, Edwards syndrome. The few trisomy-18 children who survive infancy are severely retarded. (Courtesy David D. Weaver, M.D., Indiana University School of Medicine.)

ously; perhaps only 5 percent are live-born. Their occurrence becomes more frequent with increased maternal age, as is true for all trisomies. It is interesting to note that about 80 percent of those with this syndrome are females (Figure 11—8). This may be due to a greater female viability in utero, although after birth girls with this syndrome do not survive any better than boys.

Trisomy-13 (Patau Syndrome)
The trisomy-13 syndrome is characterized by a broad nose, small cranium, small eyes that are usually nonfunctional, frequent cleft lip and palate, heart defects, and severe mental retardation (Figure 11—9). The life span is usually no more than a few weeks and the frequency is about 1 in each 15,000 live births. Children with Patau syndrome are trisomic for chromosome 13 (Figure 11—10).

Other Trisomies
A few other trisomies, including trisomy-8, have been found, but they are very rare. It is probable that nondisjunction occurs for all chromosomes, but most of the trisomies result in early abortion. Karyotypes made from aborted fetuses show trisomies for all autosomes with the

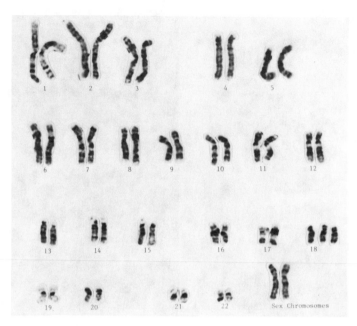

FIGURE 11–8. Karyotype of a female having trisomy-18 resulting in Edwards syndrome. Chromosomes stained using the G-banding technique. (Courtesy Catherine G. Palmer, Indiana University School of Medicine).

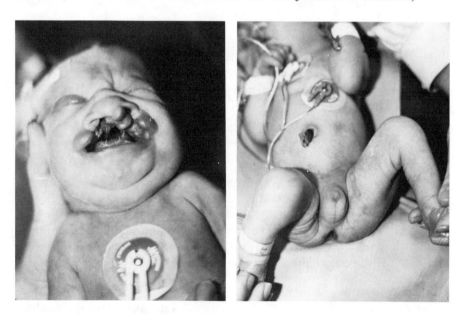

FIGURE 11–9. Newborn male with trisomy-13, Patau syndrome. Note severe cleft lip and palate and immature genitalia. (Courtesy David D. Weaver, M.D., Indiana University School of Medicine.)

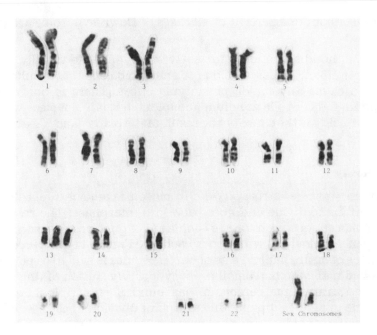

FIGURE 11–10. Karyotype of a male having trisomy-13 resulting in Patau syndrome. Chromosomes stained using the G-banding technique. (Courtesy Catherine G. Palmer, Indiana University School of Medicine.)

exception of chromosomes 1 and 5. The extent of the abnormality depends upon which genes are triplicated, although in general the greater number of genes on the longer chromosomes result in more extreme abnormalities. That this is not always true is shown by the fact that only a very few babies having trisomy-22 have been reported and these have shown extreme abnormalities. Monosomies seem to upset the gene balance so greatly that they cause death very early in embryonic development; consequently, they are not found even in early aborted fetuses. A few babies having monosomy-21 have been born alive, however.

Somatic Aneuploidy (Mosaicism)

Chromosome nondisjunction can occur in somatic cell mitosis as well as in meiosis of reproductive cells. If it occurs in an early embryo, two cell lines result, one trisomic and one monosomic. This is termed **somatic aneuploidy,** or **mosaicism.** When vital tissues are involved, abnormalities can result. If much of an important organ is monosomic, death is likely. More commonly, the trisomic cells outgrow the monosomic cells and a mosaic person ends up with most of the body cells being either trisomic or normal. Some of the less severe cases of Down syndrome result from such a mosaicism when chromosome 21 is involved. About one percent of all cases of Down syndrome are mosaics.

Somatic nondisjunction after early embryonic life usually has no noticeable effect. The small clone of aneuploid cells that results is not likely to cause any major phenotypic effect. There is one possible exception. Cancer cells are often aneuploid, but it is not clear whether the aneuploidy is the cause or the result of the malignancy (see Chapter 14).

Polyploidy

Sometimes mitosis starts and the chromosomes become duplicated, but the cell fails to divide and goes back into interphase. The result is a **tetraploid** cell, one with four of each kind of chromosome rather than the normal diploid cell with two of each (see Figure 11–11). A cell with multiples of entire haploid sets of chromosomes is said to be **polyploid**. A **triploid** man who had small jaws, syndactyly (fusion of fingers and toes), and an abnormal cerebrum, was found. Such a triploid can result when one gamete is diploid, due to lack of division in the second meiosis, and the other is the normal haploid. Triploidy has been induced in rabbits by treating the oocytes with **colchicine,** a chemical that inhibits the completion of mitosis or meiosis. Another possible cause of trip-

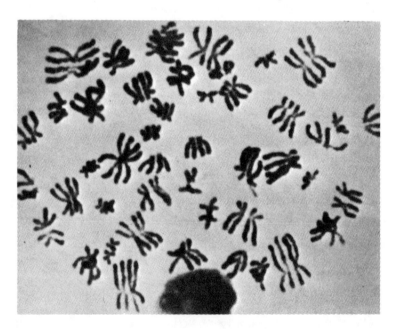

FIGURE 11–11. Tetraploid chromosomes in a cell taken from the intestine of a man who had used colchicine over a long period as a treatment for gout.

loidy is double fertilization, where two sperms enter the egg and both contribute their chromosomes to the zygote. Most triploids probably do not make it to a live birth. In fact, both triploids and tetraploids are fairly common in spontaneous abortuses.

Mosaic polyploidy is not uncommon. Some people who take colchicine as a treatment for gout have tetraploid intestinal cells. Islands of tetraploid tissue have also been found in the skin and other parts of the body, but these generally do no harm.

Aberrations Involving Chromosome Structure

Sometimes chromosomes may break apart. The portion not containing the centromere, an **acentric fragment,** then lies free without any direction during mitosis or meiosis and can be lost to the cell. Sometimes, however, it may become attached to another chromosome. Homologous chromosomes regularly break and become reattached while they are paired closely during the first meiosis. This is known as **crossing over** and, since chromosomes pair by allelic genes, it does not upset the balance of genes in the cell. It does provide for new combinations of genes on chromosomes and, therefore, for more variety. Let us see what kinds of aberrations result when a portion of a chromosome is lost or reattached in ways other than crossing over.

Deletions

A chromosome or a portion of a chromosome without a centromere is like a car without a driver. It has nothing to steer it in the proper direction in mitosis or meiosis. Hence, it will not reach the poles of the spindle in telophase and will not be included within the nucleus that is formed. Left out in the cytoplasm, it will soon be broken down into its component parts by cellular enzymes and the genes it contained will be lost to the cell. Since there are two of each kind of chromosome, however, alleles of the lost genes will be present on the homologous chromosome, but the cell will be haploid for these genes. This type of aberration is termed **deletion.** A large deletion will upset the balance of genes and cause very harmful phenotypic effects. A very small deletion may not cause any serious upset in gene balance, but it can allow recessive genes on the homologous chromosome to be expressed. A person who should be heterozygous may express a recessive because the part of the chromosome with the dominant allele has been deleted. A person with one blue eye and one brown eye, or a person who has a blue sector in an otherwise brown eye, can be produced when a small somatic deletion in early embryonic development removes the dominant allele for brown pigmentation. One eye may have descended from the cell with the deletion and will be blue, but the other eye will have been formed from cells without the deletion and will be brown.

Terminal deletion, the loss of one end of a chromosome, is the most common deletion. **Intercalary deletions** occur when a chromosome loops around itself and then the loop breaks off to form a **ring** while the ends rejoin (see Figure 11–12). If the loop includes the centromere, the ring becomes the chromosome that is retained while the two ends are deleted. If the loop does not include the centromere, the ring is deleted, leaving a shortened rodlike chromosome within the nucleus. Another type of deletion occurs when the centromere divides transversely during mitosis or meiosis. This splits the chromatids in two, so each chromosome formed includes duplicates of one end and deletions of the other end. These are known as **isochromosomes** because the portions on either side of the centromere are the same.

Translocations

Translocation involves the transfer of a portion of a chromosome to a different position on the same or a different chromosome. If it is on a different chromosome, an upset in gene balance can occur. A balanced translocation is present when a person has the deleted chromosome plus the lengthened chromosome to which the deleted piece has been attached. Segregation in meiosis, however, can result in some gametes with two sets of some genes and some gametes with none of these

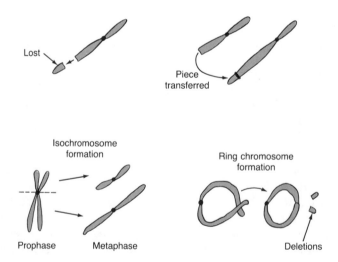

FIGURE 11–12. Kinds of chromosome aberrations that can result from breakage and reattachment of parts of chromosomes.

genes. We have already seen how this can transmit an abnormality to children when chromosome 21 is involved.

Cri du Chat (Cry of the Cat) Syndrome

Babies with the **cri du chat syndrome** (Figure 11–13) have a cry that sounds like a cat in distress because the infant's larynx or voicebox has developed improperly. Other characteristics of this aberration are a rounded face, small cranium, and severe mental retardation (see Figure 11–14). Karyotypes show that about one-half of the short arm of one of the two fifth chromosomes has been deleted (see Figure 11–15). There must be a weak place at the region of breakage because the breakage is much more frequent at this point than at any other part of the chromosome. The broken piece can be translocated onto another chromosome and normal parents with this translocation can transmit the abnormality to children through gametes that carry the chromosome 5 deletion without the other chromosome that received the deleted piece.

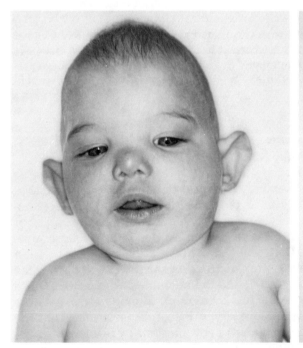 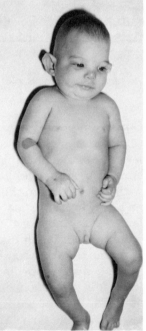

FIGURE 11–13. A child with cat-cry (cri du chat) syndrome. Note the rounded face (moon-face), the "antimongoloid" slant of the eyes, the small cranium, and the misshapen ears. Severe mental retardation is always associated with this syndrome. (Courtesy David D. Weaver, M.D., Indiana University School of Medicine.)

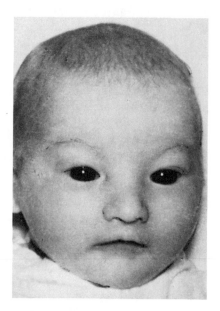 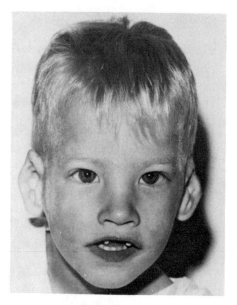

FIGURE 11–14. A patient with cri du chat syndrome. The four-year-old at right is the same child as the newborn at left. The cat cry and moon-face have disappeared, but the "antimongoloid" eye slant and misshapen ears have persisted. (Courtesy G. H. Valentine, *The Chromosome Disorders,* 3rd edition, Heineman Medical Books.)

Chronic Myeloid Leukemia

Chronic myeloid leukemia (CML) is characterized by an excess of leukocytes that have granules in the cytoplasm. It is also known as **chronic granulocytic leukemia.** The excess of leukocytes is associated with a reduction in the erythrocytes (red blood cells) and a severe anemia. However, medications are now available that reduce the output of leukocytes and permit a relatively normal life. In 1959 in a male patient in Philadelphia, CML was first shown to be associated with a shortened chromosome 22. Part of the long arm was missing. This deleted chromosome is known as the **Philadelphia chromosome.** Since then, this type of chromosome has been found in many other patients with this type of leukemia. More recent research has shown that the segment deleted from the long arm of chromosome 22 is usually translocated to the long arm of chromosome 9. Surprisingly, however, the abnormal chromosome is found only in the myeloid cells of the bone marrow, which produce the granulocytic leukocytes. CML is actually a form of cancer of these cells. It may be that these cells are particularly susceptible to agents that cause breakage at this point, or the breaks may occur in all cells but survive and reproduce only in these cells of

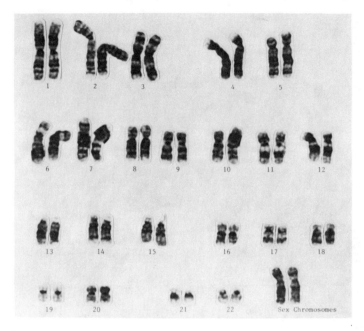

FIGURE 11—15. Karyotype of a female having cri du chat syndrome. Note that the normal number (46) of chromosomes is present, but that a portion of the short arm of one chromosome 5 has been deleted. Chromosomes stained using the G-banding technique. (Courtesy Catherine G. Palmer, Indiana University School of Medicine.)

the bone marrow. It could also be that the leukemia develops first and causes the breaks, because in those rare cases of remission the broken chromosomes no longer can be found in the marrow. At least, it is not transmissible to offspring.

Induced Chromosome Aberrations

As a rule, the same agents that cause gene mutations also cause chromosome aberrations. **Ionizing radiation,** for instance, induces both nondisjunction and chromosome breaks. Theodore Puck found that 20 percent of the cells in human tissue cultures had some kind of aberrations after exposure to 50 rads of X rays (see Figure 11—16). Less than 2 percent of the cells in control cultures had such aberrations. Regions of the world that received unusually high amounts of radioactive fallout from atmospheric tests of nuclear weapons show an increase in leukemia among the children. The carcinogenic effect of ionizing radi-

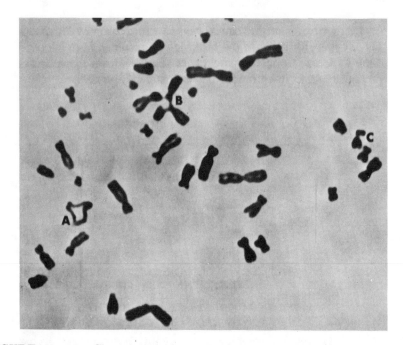

FIGURE 11–16. Chromosome aberrations in a human tissue culture cell that received 50 rads of X rays. Three aberrations (designated A, B and C) can be seen. About 20 percent of the cells in the X-ray–treated culture showed some kind of aberration, as compared to fewer than 2 percent of the cells in unirradiated controls.

ation may be due, in part, to chromosome aberrations induced in somatic cells.

LSD added to tissue cultures of human cells was found to cause chromosome aberrations in some studies. Other studies showed an increase in aberrations in the leukocytes of monkeys that had been given high doses of the drug for five weeks. One investigation showed aberrations in the leukocytes in a man who had used LSD extensively over a long period (see Figure 11–17). The frequency was as high as 25 percent as compared to less than 2 percent in those who had not used the drug. Other studies on living subjects, however, failed to show a significant increase. Differences in selection of subjects and techniques of analysis might account for these divergent results. The final conclusion must await further research and refinements of techniques.

As mentioned in Chapter 3, infection during early pregnancy with certain disease organisms can be teratogenic. In some instances it appears that chromosome aberrations are the cause of the abnormality in the fetus. A woman who has **rubella** (German measles) during early pregnancy is very likely to have a baby who is deaf, blind, or has heart

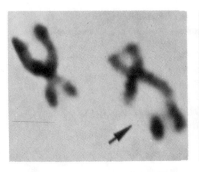 

FIGURE 11–17. Chromosome aberrations found in a cell of a man who had used LSD extensively over a long period of time. A deletion is shown at left, and isochromosome formation at right. Over 20 percent of his leukocytes had such aberrations. (Courtesy E. J. Egozcue.)

abnormalities. The type of abnormality depends upon the age of the fetus when the infection occurs. Seven to twelve weeks is the period during which the organs are in critical stages of development and most susceptible to damage. Cultures from the blood cells of persons in the third day of infection with rubella revealed chromosome aberrations in 32 to 70 percent of the leukocytes. Infection with the **herpes virus, type II,** which has expanded from a very rare infection to one of the most common venereal diseases, is also teratogenic during early pregnancy, possibly because it induces chromosome aberrations. There is even some evidence that the influenza virus can have a similar effect.

Symbols for Chromosome Aberrations

As new methods of chromosome analysis revealed many types of aberrations, it was necessary to establish standardized symbols to represent the various chromosome complements and any aberrations that might be present. Such a method was established at a Paris conference in 1971. A normal male is designated as 46, XY (a total chromosome number of 46, with X and Y as the sex chromosomes). A male with Down syndrome would be 47, XY, + 21 (there is an extra chromosome 21). For aberrations involving parts of chromosomes, the number of the chromosome with the aberration is given, then the arm involved (p or q), and then a plus or minus sign to indicate the addition or loss of a part of this arm. A female with a deletion on the short arm of chromosome 8 would be 46, XX, $8p^-$. A translocation of a part of the long arm of chromosome 6 to the short arm of chromosome 16 would be 46, XX, $t(6q^-, 16p^+)$. Some other representations are shown in Table 11–1.

TABLE 11–1   Symbols for various chromosome complements

Symbol	Condition
46, XY	normal male
46, XX	normal female
45, X	female with Turner syndrome
47, XY, +18	male with trisomy-18 (Edwards syndrome)
47, XYY	male with two Y chromosomes
45, XX, −8	female with monosomy-8
46, XX, t(5p⁻, 13q⁺)	female with translocation of part of short arm of 5 to long arm of 13
45, X/47, XXX	female mosaic, some cells with one X and others with triple-X
46, XY, i(18q)	male with isochromosome having two long arms of 18 and no short arms of 18
46, XX, r(15q)	female with long arm of 15 in form of a ring

PROBLEMS AND QUESTIONS

1   If $2n$ is used to symbolize a normal diploid individual and $2n + 1$ to symbolize a trisomic, what symbols would be appropriate for a
  (a) monosomic?
  (b) triploid?
  (c) tetraploid?

2   How many chromosomes would humans having the following conditions be expected to have in each of their body cells?
  (a) trisomic
  (b) monosomic
  (c) triploid
  (d) tetraploid

3   Using the system of symbols shown in Table 11–1, write the symbol appropriate for each of the following individuals:
  (a) male with classical Klinefelter syndrome
  (b) female with cri du chat syndrome
  (c) female with Patau syndrome
  (d) male with the classical form of Down syndrome
  (e) phenotypically normal female carrier of the translocation of the long arm of chromosome 21 to the long arm of chromosome 14

4   What is the special value of chromosome banding techniques for detecting structural chromosome aberrations such as deletions and translocations?

5   Males with trisomy-21 Down syndrome are sterile but females with trisomy-21 have been known to reproduce. A person who is trisomic for an autosome should produce two kinds of gametes with equal frequency—normal gametes and ones with an extra chromo-

some. If a trisomy-21 female produces children by a normal male, what is the probability that the offspring will have Down syndrome? What is the probability that the offspring will be a female who does *not* have Down syndrome?

6 If ABCD·EFGH represents one normal chromosome and JKL·MNOPQ represents another normal chromosome (where the · represents the centromere), what structural chromosome aberration is represented by each of the following?

   (a) ABCD·EF
   (b) AD·EFGH
   (c) QPONM·MNOPQ
   (d) ABCD·ENOPQ
   (e) GH
   (f) JKL·MPONQ

7 List the three types of Down syndrome mentioned in this chapter. Which type is the most common? the least common? the least severe? Which type is most likely to recur among siblings of an affected child?

8 A twenty-five-year-old woman (whose husband was twenty-six) gave birth to a Down syndrome child as her first offspring. The attending physician recommended that karyotypes be prepared for the woman and her husband, as well as for the child, before the couple decides on having other children. What do you perceive to be the purpose of the physician's recommendation?

REFERENCES

BORGAONKAR, D. S. 1977. *Chromosomal Variation in Man: A Catalog of Chromosomal Variants and Anomalies,* 2d ed. Alan R. Liss, Inc. New York.

DEGROUCHY, J., and C. TURLEAU. 1977. *Clinical Atlas of Human Chromosomes.* John Wiley & Sons. New York.

DUTRILLAUX, B., and J. LEJEUNE. 1975. New techniques in the study of human chromosomes: methods and applications. *Advances in Human Genetics* 5:119–56.

FUCHS, F. 1980. Genetic amniocentesis. *Scientific American* 242(6):47–53.

HOLMES, L. 1978. How fathers can cause the Down syndrome. *Human Nature* 1(10):70–72.

SMITH, D. W., and A. A. WILSON. 1973. *The Child with Down's Syndrome (Mongolism).* W. B. Saunders Co., Philadelphia.

THERMAN, E. 1980. *Human Chromosomes: Structure, Behavior, Effects.* Springer-Verlag. New York.

YUNIS, J. J., Editor. 1977. *New Chromosomal Syndromes.* Academic Press. New York.

twelve

Somatic Cell Genetics and Genetic Engineering

IN THE PAST MOST GENETIC STUDIES HAVE concentrated on the assortment of genes during gamete formation, the ratio of genotypes and phenotypes in zygote formation, and the action of genes in the formation of the body during embryonic development. We should remember, however, that many genes continue to function in the body cells throughout life. These genes produce enzymes, hormones, and other products that are vital to everyday existence. More and more attention is now being given to the study of the functioning of these genes in somatic cells. Many new techniques now make possible a detailed analysis of these functions. Surprisingly, we can study this activity better when the cells are removed from the body and grown in culture dishes (in vitro) much as we have been growing bacteria for a long time.

The perfection of these somatic cell techniques has made possible the manipulation of the hereditary material in ways beyond the wildest dreams of geneticists during the first three-fourths of this century. A new field of genetics, known as **genetic engineering,** has arisen and is very much in the news today because of the great potentialities it offers, as well as the problems it has created. Much has already been done and much, no doubt, will be done in the future to improve the quality of human life through genetic engineering. At the same time, however, controversy has arisen over the many legal and moral questions involved as we create new gene combinations in ways that were never envisioned when our legal and moral codes were formulated. In this chapter we shall consider some of the methods of somatic cell genetics and how they are being used in genetic engineering, as well as the problems that have arisen.

Tissue Culture (Growing Cells Outside the Body)

The discovery of satisfactory methods of growing cells outside the body was a necessary prelude to most research in the field of genetic engineering. Most somatic cell studies must be done on tissue culture cells because it is not possible to use the proper techniques on cells while they are still a part of a living body. Fortunately, many types of human cells can be grown in a culture if they are provided with the same conditions, as closely as possible, as are found in the body. Nutritional requirements must be met, and the proper temperature and a sterile environment maintained.

Advantages of Cultured Cells

Cultured cells can be experimented with in ways that would be unthinkable if they were still a part of a human body. For instance, they can be exposed to high-energy radiation and harsh chemicals that would be very harmful to a person. A good example of how somatic cells might be used to increase our understanding of human genetic defects is the study of **familial hypercholesterolemia,** an excess of cholesterol in the blood, which leads to early on-set coronary heart disease. Experiments with cells from persons who have inherited this condition have shown that the defect is the result of genes that affect the cell surface binding and phagocytosis (engulfing) of cholesterol. The result is an excessive level of cholesterol in the blood because it is not controlled by natural limitation reactions. Scientists have obtained this information by exposing cells to high concentrations of cholesterol, which would not be possible in a living body.

Kinds of Cells That Can Be Cultured

Not all kinds of human cells can be grown in a culture outside the body. Some, such as the cells of nerves and muscles, are so highly differentiated that they will not grow and divide in culture using our present techniques. Cells that grow and divide regularly in the body are the ones that are most likely to be grown successfully in culture. The cells most commonly used are discussed below.

Fibroblasts   **Fibroblasts** grow very well in culture and are widely used for genetic investigations. They are generalized cells that have not become highly differentiated. They are widely distributed in many tissues, and if some of these tissues are placed in a culture the fibroblasts will survive and grow while most of the other types of cells will die. They have one important drawback, however. After they have undergone up to about 200 divisions they become senescent and growth ceases. Cells taken from young donors live longer than those from older donors. The exact reason for this has not been determined, but it is almost certainly related to the natural aging of human beings. Further study of the aging process might enable us to delay senescence in older people and prolong the period of vital activity in life.

Senescence of cultured cells can be prevented if they are infected with certain types of benign viruses. The virus does not harm the cells, but causes them to form a permanent line that can be kept forever, apparently, as long as they are subcultured at regular intervals and the proper conditions are provided. Or, they can be frozen and reactivated at any time for further study. Also, the virus infection will cause the cells to lose what is known as **contact inhibition.** Without the infection the cells will grow until they contact other cells, then growth stops. After the infection, however, they will continue growing and grow over the other cells. This is a good quality when a large number of cells is needed for experimentation.

Malignant Cells   **Malignant** (cancerous) **cells** have lost contact inhibition and they do not become senescent. Hence, they can be cultured for long periods of time. One cell line from a cervical cancer of a woman named Henrietta Lacks has been cultured since it was isolated in 1951. It is known as the **HeLa strain** and is still being widely used all over the world today. In fact, because of its vigor, the HeLa strain has become a real nuisance as a contaminant of other cell lines that researchers attempt to culture.

Lymphocytes   It is also possible to culture certain types of white blood cells. These are the **lymphocytes,** leukocytes with a single large nucleus, that play such an important role in the immune reaction (see Chapter 7). While these cells do not normally grow and divide in the blood stream, they can be induced to grow and divide in cultures if they are treated with an extract of the red kidney bean, known as phytohemagglutinin (PHA). This extract has long been used to agglutinate red blood cells so they can be easily removed from whole blood when only white blood cells are desired. Treated lymphocytes are the cells most commonly used for studies of human chromosomes, as we learned in Chapter 1.

Mutation Identification in Somatic Cell Cultures

Cultured somatic cells can be studied to detect chromosome aberrations by cytological examination, and they can also be used to detect somatic mutations and establish the mutation rate for specific genes. Two primary methods, described below, are used to isolate and identify mutations in cultured cells.

Use of Selective Media

Identification of somatic cells with new mutations can be done through the use of selective media. A **selective medium** is one that allows only the cells with a certain phenotype to survive. The technique has long been used to detect mutations in bacteria. For instance, to detect a mutation of a certain bacterium to penicillin resistance, the cells are placed on a medium containing penicillin. Only those mutants that can resist the antimetabolic action of this antibiotic will grow. Hence, it is easy to isolate mutants for penicillin resistance even though they may occur only once in several million cell generations.

Bacteria that can grow on a medium containing penicillin have a mutant gene that can produce an enzyme that neutralizes the antibiotic. Some selective media used for mammalian cells also contain toxic substances that can be tolerated only by mutant cells that have enzymes that can neutralize them. Or, the media may contain substances that become changed into toxic substances if a certain enzyme is present and only the mutant cells that do not have this enzyme can survive. Still other selective media may be deficient in some vital amino acid or vitamin, and the only cells that can survive are those that produce enzymes that can synthesize these products from other substances in the medium.

Use of Electrophoretic Separation of Proteins

Proteins that may be of the same basic structure but differ slightly in even one of their amino acids can often be separated and identified by **electrophoresis**. Amino acids may have a positive charge, a negative charge, or no charge, and variation in these amino acids will cause entire proteins to migrate at different rates when exposed to an electrical field. This is the method used in separating the many variant forms of hemoglobin as was described in Chapter 10. It can also be used to identify mutants in cell cultures. As an example, mouse glucose-6-phosphate dehydrogenase (G6PD) can be distinguished from human G6PD by this technique. In hybrid cells where both types of G6PD are present, the gel will have multiple bands. Hence, any hybrid cell that has been shown to contain human G6PD will contain the human X chromosome since this enzyme is controlled by an X-linked mutant gene. You will recall that persons who are deficient in this enzyme suffer a severe anemia after eating certain types of beans and after taking certain drugs (Chapter 6).

Determining the Rate of Mutation in Somatic Cells

Mutations to azaguanine resistance in human somatic cells can be isolated by use of a selective medium. **Azaguanine** is chemically similar to guanine, one of the bases of DNA. When azaguanine is present, normal cells will take it up and incorporate it into their DNA during replication. The DNA that includes this antimetabolite, however, does not function properly and cells containing it will die. A certain dominant gene codes the production of an enzyme that is necessary for the uptake of azaguanine, but its recessive allele does not produce the enzyme in a functional form. This enzyme is hypoxanthine-guanine phosphoribosyltransferase, most often abbreviated as HGPRT. It is the enzyme that is deficient in persons with Lesch-Nyhan syndrome, a serious defect characterized by the production of an excess of uric acid (see Chapter 6). Heterozygous normal cells are treated with high-energy radiation or chemical mutagens and then placed on a selective medium containing azaguanine. A mutation of the dominant gene to its recessive allele results in a homozygous recessive cell that cannot produce the enzyme HGPRT and cannot substitute azaguanine for the normal guanine. Only this mutant cell will grow on the selective medium. By comparing the number of cells placed on the medium with the number of cells that survive and grow, the rate of mutation to azaguanine resistance can be determined.

When the cells are exposed to a maximum intensity of mutagens (an intensity just short of a lethal dose) they have a mutation rate to azaguanine resistance of about 10^{-4} (1 in 10,000 cells). The spontaneous mutation rate has been established at about 10^{-8} by placing untreated cells on the selective medium. These figures are valuable in establishing the relative mutation rate of different genes, but they will be lower than the rate established by surveys of populations. The rate obtained from cultures of somatic cells is per cell generation, while the population rate is per generation of people. A single mutation in a spermatogonial cell, for instance, might be replicated many times and appear in many sperm, thus greatly increasing the chance that it will be transmitted to any one child. The best evidence we have for the frequency of this particular mutation in populations is about 10^{-5}, about 1,000 times more frequent than the rate found in cultured cells. We do not purposely expose humans to high concentrations of mutagenic agents in order to detect the induced mutation rate, but we would expect a similar higher rate to appear in children of exposed parents when compared with the induced rate in cell culture.

Prenatal Detection of Mutant Genes

It is also possible to use the selective media technique to identify certain metabolic disorders prenatally. Some of the amniotic fluid removed by amniocentesis can be placed on the selective medium, and if the cells in this fluid grow, we know that the enzyme being investigated is not present and a disease, such as Lesch-Nyhan syndrome, will be expected if the pregnancy goes to full term. Also, electrophoretic separation of the proteins in cells from the amniotic fluid will indicate the presence of certain types of mutants.

Hybridization of Somatic Cells

When cells from two different species of mammals are grown together in the same culture, some of them will fuse with one another to give tetraploid cells with the full diploid number of chromosomes of both species. This was an amazing discovery because attempts at hybridization of gametes of unrelated species have been unsuccessful. Gametes of closely related species or even closely related genera on occasion will fuse and give a hybrid, but the physiological differences between the gametes of widely different species will not allow a union. Somatic cells, however, even of such divergent species as the mouse and the human, will fuse when cultured together. Before you have visions of some mad scientist implanting a fused mouse-human cell into a woman to give a hybrid monster, it should be pointed out that the hybrid cells will gradually lose the human chromosomes during repeated divisions. The mouse mitosis seems to be completed in a somewhat shorter time period than the human mitosis and as the nuclear membrane is formed in telophase the more slowly migrating human chromosomes are often excluded from the nucleus. This feature is valuable for genetic studies as we shall learn shortly (Figure 12–1).

Increasing the Rate of Fusion

Spontaneous fusion of cells of different species is not frequent and some method was needed to increase the rate. This was found when the cells were treated with **Sendai virus** particles that had been inactivated with ultraviolet rays. This virus causes no harm to the cells, but greatly increases the frequency of fusion and induces cells of widely different species to fuse. The chemical **polyethylene glycol** has a similar effect and is used most commonly today for increasing the fusion rate.

Isolating Fused Cells

The next problem involved the isolation of the fused cells from those that had not fused. This was accomplished by use of selective media as has been described for detection of mutations in cultures. For instance,

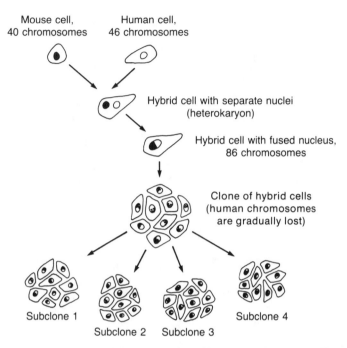

Mouse cell, 40 chromosomes

Human cell, 46 chromosomes

Hybrid cell with separate nuclei (heterokaryon)

Hybrid cell with fused nucleus, 86 chromosomes

Clone of hybrid cells (human chromosomes are gradually lost)

Subclone 1

Subclone 2

Subclone 3

Subclone 4

FIGURE 12–1. Somatic cell hybridization. Human and mouse cells cultured together can be induced to fuse by treatments with inactivated Sendai virus or with polyethylene glycol. As the hybrid cells divide, human chromosomes tend to be preferentially lost, producing subclones containing different human chromosomes.

suppose that mouse cells require substance A, but not substance B, in the medium in order to survive, while human cells do not need A, but do need substance B. If a mixed culture is placed on a medium lacking both A and B then only the hybrid cells will survive. The surviving cells can then be transferred to a fresh culture medium to give a pure culture of hybrid cells (Figure 12–2).

Assignment of Genes to Chromosomes

The fact that human chromosomes are gradually lost as mouse-human hybrid cells continue to divide makes possible the assignment of specific human genes to their proper chromosomes. The first assignment using this technique was for the gene that produces the enzyme **thymidine kinase** (TK). Humans have a gene for the production of this enzyme, but certain mouse strains do not. A medium was devised with certain chemicals that would allow survival only of cells that could produce the TK enzyme. After a number of divisions of the cells on this medium, the surviving cells were examined to determine which human chromosomes remained. Some of these cells proved to have only one human chromosome, others had several human chromosomes, but all

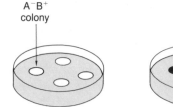

I. Substance A in medium

II. Substance B in medium

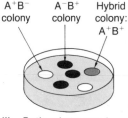

III. Both substances A and B in medium

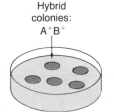

IV. Neither substance A nor substance B in medium

FIGURE 12–2. Use of selective medium to isolate somatic cell hybrids. See text for further details.

of them had chromosome 17 (see Table 12–1). Hence, it was concluded that this was the chromosome on which the gene for the enzyme was located. Since this initial assignment by Weiss and Green, in 1967, many other human genes have been assigned to particular chromosomes by this method. For further details, see Figure 12–3 and the reference by McKusick and Ruddle cited at the end of the chapter.

Malignant Cell Investigations

We have obtained important information about malignant cells by cell hybridization. Fusion of normal human cells with malignant human cells yields hybrid cells that do not show malignant characteristics unless and until they lose a specific chromosome that came from the normal cells. Hence, it appears that normal cells have a gene, or genes, on this chromosome that regulates the rate of cell growth and division. When malignant human cells are fused with mouse cells, however, the hybrids express the malignant characteristics and will grow as malignancies when injected into living mice. Hence, it appears that mouse cells lack the genes needed for normal growth and division of human cells.

Further interesting studies have shown the location of areas on human chromosomes that can integrate a cancer-causing virus. **Sim-**

TABLE 12–1 Identification of the chromosomal location of the gene for thymidine kinase (TK).

Human cells that produced thymidine kinase (= TK$^+$) were fused with mouse cells that did not produce the enzyme (= TK$^-$). Only those hybrid cells containing human chromosome 17 were capable of producing thymidine kinase, thus suggesting that the TK gene is located on chromosome 17.

Mouse/Human Clone	Human Chromosomes Persisting in the Clone	Presence of Thymidine Kinase*
A	5, 9, 12, 21	−
B	3, 4, 17, 21	+
C	5, 6, 14, 17, 22	+
D	3, 4, 9, 18, 22	−
E	1, 2, 6, 7, 20	−
F	1, 9, 17, 18, 20	+

* − = TK absent; + = TK present

ian virus 40 (SV40) does not kill cells, but will cause normal cells in culture to become malignant. These malignant cells can then be fused with mouse cells and, after allowing time for the elimination of most of the human chromosomes, they can be tested for the presence of the SV40 virus by exposure to antibodies that will react with the antigen of this virus. It was found that only the hybrid cells that still contained human chromosome 7 would react with the antibodies; hence, this must be the only chromosome into which the virus can become integrated.

Determining Gene Position on the Chromosome

While hybridization of somatic cells allows us to determine which chromosome carries a particular gene, it does not give us information about the specific location of the gene on the chromosome. One way in which this can be determined, however, is by using another characteristic of somatic cells: They will take up pieces of chromosomes by phagocytosis, an engulfing process, and incorporate these pieces into their own chromosome complex.

Human chromosomes can be broken up in a blender or by the use of restriction enzymes that cleave the chromosomes at particular points. When these pieces of chromosomes are added to a culture of mouse cells, some of the pieces will be taken up and incorporated into the mouse cell nuclei. By the use of selective media we can determine which human genes these mouse cells now express. Then, by cytological studies of the banding patterns we can recognize which piece of

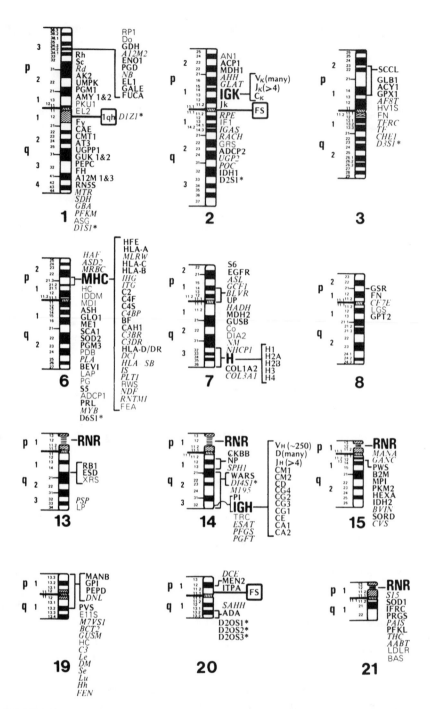

FIGURE 12–3. Map of human chromosomes showing locations of numerous genes, many assigned on the basis of somatic cell hybridization studies. Only a few genes discussed in this book are specifically identified here. Chromosome 1—Rh blood group (Rh), phenylketonuria (PKU); chromosome 4—MNSs blood groups; chromosome 9—ABO blood groups; chromosome 11—β chain of hemoglobin (HBB); chromosome 15—hexosaminidase A (Hex A) (Tay-Sachs

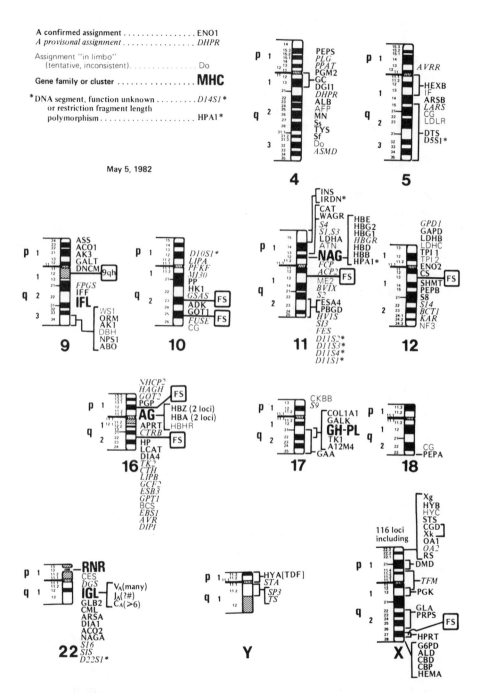

A confirmed assignment ENO1
A provisional assignment . . . . *DHPR*

Assignment "in limbo"
(tentative, inconsistent). . . . . Do

Gene family or cluster . . . . **MHC**

*DNA segment, function unknown . . . . *D14S1* *
or restriction fragment length
polymorphism HPA1*

May 5, 1982

disease); chromosome 16—α chain of hemoglobin (HBA); chromosome 17—thymidine kinase (TK1); Y chromosome—Y histocompatibility antigen (H-Y); X chromosome—Xg blood group, protan (CBP) and deutan (CBD) colorblindness loci, hemophilia A (HemA). (Courtesy Victor McKusick, Johns Hopkins University.)

207

which chromosome has been incorporated into the mouse cells. This will tell us the particular part of the chromosome that contains the genes that are expressed. It can also show us which genes are closely linked. If several human genes are expressed and only a small part of a single human chromosome is present, then these genes must lie close together on this part of the chromosome.

This transfer of small blocks of genes offers great possibilities for the correction of genetic defects in living organisms. We have already used it to transfer human genes into living mice. Cultured mouse somatic cells that have taken up pieces of human chromosomes can be injected into the blastocyst (early embryo) of mice. The result will be a mouse with two cell lines, one of which includes human genes. Some of the gametes of these mice will have the human genes, and offspring produced from these gametes will have the human genes in all their cells. We can then conduct experiments on these mice to study the action of certain human genes in ways that would not be possible in a human body.

Some attempts have been made to introduce normal genes into cells taken from a person with defective genes for certain enzymes or hormones and then reintroduce the transformed cells back into the donor with the hope that the normal genes will take over and produce the products for which the donor is deficient. For instance, cells from the liver of a person with phenylketonuria (PKU) might be cultured. Then pieces of chromosomes from normal (not having PKU) persons might be added to the cultures and some of the pieces with the gene for normal phenylalanine hydroxylase production would be taken up and incorporated into the chromosomes of the cells. Some of these cells could then be reintroduced into the liver of the PKU victim with the hope that they could then begin normal production of the missing enzyme phenylalanine hydroxylase. So far such attempts have not been successful, but with refinement of techniques we can hope for success in the future.

Complementation Tests for Allelism

Sometimes it is difficult to determine if two recessive genes that cause similar, yet somewhat different, effects are the result of alleles of the same gene locus or if they are the result of nonallelic genes at different loci. In experimental animals it is easy; we simply cross an animal that expresses one of the characteristics with one that expresses the similar characteristic. If normal offspring result, the genes appear to be nonallelic; these offspring would be heterozygous for both genes involved and would express the dominant allele of each. If the genes for the recessive characteristics are alleles, however, the phenotype of the offspring will not be normal; they would be heterozygous for the two recessive genes and no normal, dominant allele would be present. The phenotype would most likely be in between the phenotypes produced by the two recessive genes.

Such an efficient method, however, cannot be used for humans because we cannot breed people in order to obtain genetic information, but we can obtain the information from somatic cells. We do this by the **complementation test,** which consists of the cultivation of two human cell lines in the same culture dish. When these cultures reach such a high density that the cells are in close contact, they will exchange metabolic products. A cell may be missing a product, but it can be obtained from one of its neighbor cells from a different line that may have that product.

As an example of how this principle can be used to determine allelism, let us assume that one cell line is homozygous for a recessive gene c and cannot produce product C, but it has the dominant gene D and can produce product D. The other line is homozygous for the recessive gene d and cannot produce product D, but it has the dominant gene C and produces product C. When the two cell lines grow in close proximity, both will contain products C and D. Each has absorbed the missing product from the other, a process known as complementation. Such a reaction indicates that the genes c and d are not alleles; there can be no complementation if the genes are alleles. If one cell line is homozygous for the recessive gene e, and the other is homozygous for e^a, then neither can produce the normal product E. Hence, the cell lines will both remain defective with respect to this product.

Using this technique, it was possible to show that **Hunter** and **Hurler syndromes** are caused by nonallelic genes. Both of these syndromes are the result of defects in metabolism of mucopolysaccharides; both are characterized by mental retardation, dwarfism, and grotesque facial features, but there are differences between the two. Cells from individuals with either of these syndromes when grown in culture have metachromatic granules in the cytoplasm. When they are grown together in close contact with one another, however, no such granules form. Hence, it is evident that each cell line was supplying something that was missing in the other cell line, so the defects must be due to genes at different loci. We have since learned that the Hurler syndrome results from a recessive autosomal gene, while the Hunter syndrome is the result of a recessive gene on the X chromosome.

Persons with **xeroderma pigmentosum** have skin that is extremely sensitive to sunlight. Even small exposures result in the development of large freckles that often become cancerous. The condition is the result of a defect in the DNA repair mechanism that protects most people from this type of damage from the ultraviolet rays of sunlight.

When cells from different people who had xeroderma pigmentosum were cultured together, normal growth occurred in many cases. This indicates that the abnormal response to ultraviolet rays can result from a number of genes at different loci, and that the DNA repair mechnaism is dependent on the normal functioning of all of these genes.

Gene Splicing (Recombinant DNA Techniques)

Gene splicing, which results in recombinant DNA, is one of the most exciting aspects of genetic engineering. The DNA molecule can be cut at precise points and specific genes removed and transferred from one cell to another, even from such widely divergent species as human beings and bacteria (see Figure 12–4). Let us see what has been done in this field and what we might expect in the future.

Present Accomplishments

The gene for the **insulin** molecule has been transferred from human cells into bacteria and these bacteria are now producing insulin in commercial quantities. Tests of this insulin's effectiveness with human patients have shown that it can control insulin-dependent diabetes. This insulin has advantages over that from the pancreases of slaughterhouse animals, the only source in the past. The insulin from the

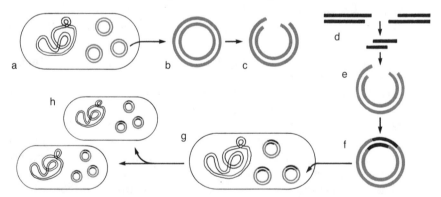

FIGURE 12–4. The mechanics of producing recombinant DNA in the bacterium, *Escherichia coli*. (a) Each *E. coli* cell contains one chromosome and several small circlets of DNA called plasmids. (b) The plasmids can be isolated from bacteria that have been ruptured. (c) Restriction enzyme can break the plasmid at specific locations. (d) The same restriction enzyme can be used to remove a segment of DNA—say the insulin gene—from a human cell. (e) The human gene is spliced into the *E. coli* plasmid and the result (f) is "recombinant" DNA. (g) The recombinant plasmids can now be reinserted into *E. coli* cells where they will subsequently be reproduced (h) each time the bacterium divides.

bacteria is produced by human genes and has a molecular structure identical to that produced in the human body. That from other animals is slightly different and some diabetics cannot tolerate this foreign type of insulin. Recombinant DNA technology has also enabled bacteria to receive human genes for the production of **interferon,** that substance produced by human cells in response to virus infections and which protects other cells from the infection. Interferon holds great promise in the treatment of virus diseases, such as influenza and hepatitis, and preliminary tests at the M. D. Anderson Hospital in Houston have shown it to be of value in treating cancer, especially of the breast and lymph nodes. In the past, research on interferon has been hampered by its scarcity and expense. It has been obtained from human lymphocytes, primarily after these cells were infected by a virus in a tissue culture. The yield, however, has been very low and the interferon very difficult to obtain. Now bacteria can produce human interferon and both the problem of quantity and price should be solved.

Human **growth hormone** is another product that has been very scarce and expensive. In the past it had to be obtained from the pituitary glands of humans who had just died, and not many of these are available. Hence, many children afflicted with dwarfism, who could have benefited from this hormone, had to be denied because of its scarcity and cost. Tests at the Ormond Hospital in London have shown that the hormone produced by bacteria is just as effective as the one from human glands.

Prospects for the Future

The prospects for the use of recombinant DNA techniques for the benefit of humanity are enormous. They include the introduction into bacteria of the genes that are able to break down oil slicks on water, those to produce fertilizers that are now becoming in short supply in the world, and those to enable the bacteria to capture the energy of sunlight for use as fuel or food. The prospects also include introducing genes for nitrogen fixation into plants that do not now have this ability. The legumes (beans and relatives) can be infected with bacteria that have the genes that enable them to combine nitrogen from the air with oxygen to form nitrates that are needed by all plants for growth. If these bacterial genes can be transferred to corn, wheat, rice, and other food plants, the increase in world food supply would be dramatic. Also, as we previously mentioned in this chapter, we might be able to transfer genes that are normal into people who suffer from deficiencies and thus overcome many genetic diseases. Progress has been made in research to overcome hemophilia and sickle-cell anemia by these techniques. If cloning in animals is expanded (it has now been done in mice), we could even produce entire human beings from tissue culture cells containing recombinant DNA, perhaps including some genes from bacteria or other forms of life. Some of these predicted accomplishments may have already been done by the time this book is in print, judging by the speed with which discoveries are being made in recent years.

The Techniques of Gene Splicing

How can molecular geneticists cut up the DNA molecule at precise points so as to remove specific genes from one cell and transfer them into the gene complex of another cell? This seemingly impossible task is accomplished by means of enzymes, known as **restriction enzymes** or **nucleases,** that cause a break in the DNA chain only where specific bases are adjacent. The pieces of DNA can then be placed in bacterial cultures or tissue cultures of the cells of multicellular organisms where they will be taken up and become a part of the genetic constitution of those cells. Selective media can then be used to isolate the particular cells that have taken up the desired genes.

Problems Arising from Gene Splicing

The many prospects for good that are possible with gene splicing are tempered with possible dangers. Some of the bacteria being used are common harmless organisms, such as *Escherichia coli* that inhabits the colon of all people. Some fear that with the introduction of new genes into these bacteria some new strain may be created that will be virulent and cause great epidemics, against which we have no natural immunity. As a result, a moratorium was temporarily declared on recombinant DNA research involving certain organisms, but the research was later resumed under strict regulations to minimize any risks. Then too, there are moral considerations. As we experiment with human cells, we may put together various gene combinations that would result in some sort of super human if these cells were to be cloned. Many feel that we do not have the right to so tamper with the natural process of gene recombination and reproduction in humans or in other organisms.

At any rate, we can be sure that genetic engineering will be of importance in our future. No longer must we be content with the assortments of genes that have accumulated naturally, but can boldly rearrange the genetic constitutions in ways that suit our needs and desires.

PROBLEMS AND QUESTIONS

1 What is meant by *contact inhibition* and why is it an important part of the controlled growth of new tissue when injuries are being repaired?

2   What advantages for genetic studies do cultured human cells have over cells in the body?

3   Malignant cells were among the first to be cultured and maintained for long periods in culture. More recently other cells have been widely used. Explain.

4   Why is it easier to identify mutations in humans through the use of cultured cells than in pedigree studies of families?

5   How can mutations with a harmful effect be detected during early fetal development?

6   Some people, upon learning of the success of fusion of somatic cells of different species, have expressed fear that a hybrid human-mouse, or other such weird combination, might be produced. What are the chances for such hybridization?

7   How have somatic cell fusion studies indicated a genetic background for malignancies?

8   The determination of the location of genes on human chromosomes could not have been done before the perfection of banding techniques for studying human chromosomes. Explain why.

9   When two different lines of cells are grown in close proximity they may exchange cell products. How can this principle be used to determine if two recessive genes are alleles?

10   How might our discoveries on the intake of pieces of chromosomes by somatic cells be useful in correcting certain human genetic defects?

11   The possible financial rewards from recombinant DNA research have become so great that a number of corporations have been established to try to make combinations that will be of great economic value to them. Will such a trend probably accelerate or hinder the progress of discoveries in this field? Explain your answer.

12   What are the advantages of obtaining hormones from bacteria that have received human genes over obtaining the hormones from other mammals?

REFERENCES

ANDERSON, W. F. and E. G. DIACUMAKOS. 1981. Genetic engineering in mammalian cells. *Scientific American* 245(1):106–21.

ANDERSON, W. F. and J. C. FLETCHER. 1980. Gene therapy in human beings: when is it ethical to begin? *New England Journal of Medicine* 303(22):1293–97.

DAVIS, B. D. 1977. The recombinant DNA scenarios: Andromeda strain, chimera, and Golem. *American Scientist* 65(5):547–55.

EPHRUSSI, B. 1972. *Hybridization of Somatic Cells.* Princeton University Press. Princeton, New Jersey.

EPHRUSSI, B., and M. C. WEISS. 1969. Hybrid somatic cells. *Scientific American* 220(4):26–35.

FREIFELDER, D., Editor. 1978. *Recombinant DNA.* Readings from *Scientific American.* W. H. Freeman and Co. San Francisco.

HARRIS, H. 1970. *Cell Fusion.* Harvard University Press. Cambridge, Massachusetts.

McKUSICK, V. A. 1980. The anatomy of the human genome. *The Journal of Heredity* 71(6):370–91.

McKUSICK, V. A., and F. H. RUDDLE. 1977. The status of the gene map of the human chromosomes. *Science* 196:390–405.

NATIONAL ACADEMY OF SCIENCES BULLETIN. 1977. *Research with Recombinant DNA.* Washington, D.C.

PUCK, T. T. 1972. *The Mammalian Cell as a Microorganism: Genetic and Biochemical Studies in Vitro.* Holden-Day. San Francisco.

RINGERTZ, N. R., and R. E. SAVAGE. 1976. *Cell Hybrids.* Academic Press. New York.

RINGERTZ, N. R., and T. EGE. 1977. Use of mutant, hybrid, and reconstructed cells in somatic cell genetics. *Genetic Mechanisms of Cells. Cell Biology: A Comprehensive Treatise,* Vol. 1 (Ed. by L. Goldstein and D. M. Prescott). Academic Press. New York.

RUDDLE, F. H., and R. S. KUCHERLAPATI. 1974. Hybrid cells and human genes. *Scientific American* 231(1):36–44.

TIME. 1980. Interferon: The IF drug for cancer. *Time* (March 31, 1980): 60–66.

————. 1981. Shaping life in the lab: The boom in genetic engineering. *Time* (March 7, 1981):50–59.

WADE, N. 1979. *The Ultimate Experiment: Man-Made Evolution,* 2d ed. Walker and Co. New York, New York.

thirteen

Behavior Genetics

DOES HEREDITY PLAY A PART IN YOUR BE-
havior? Is most of your behavior learned or do
your genes strongly influence the way you react to
your environment? There is no doubt that very
complex behavioral patterns can be inherited. In
many animal species heredity is the predominant
force. Insects, for instance, are hatched with about
all the reactive potentials they will ever have or
need. Practically all of their reactions are instinctive and cannot be altered by training. A moth will
continue to fly toward a flame even after its wings
have been singed by heat.

Vertebrate animals have the power to modify
their reactions by learning, but most of them still
depend on inheritance as the primary directing
force behind their reactions. Birds, for instance,
react to certain hormonal changes by building
nests of the same materials and in the same way as
has been done by their ancestors even though they
may never have seen such a nest. Salmon migrate
from the ocean and go up rivers to lay their eggs in
a particular spring-fed stream. In an attempt to
reach that stream, they will beat themselves to
death against barriers, such as dams, that may bar
their way. In mammals, learning plays a more
important role and some instinctive patterns can
be altered by training. Some breeds of dogs have
an instinctive urge to chase chickens, but stern
discipline can overcome this reaction. Horses inherit a certain gait for running, although some
training is needed to reinforce this pattern for pac-
ers and trotters in harness racing. The subhuman
primates have a great capacity for learning, as evi-
denced by the complex reactions, even including a
rather extensive sign language, that have been
taught to chimpanzees.

With respect to humans, however, the picture is
not so clear. We certainly have a greater capacity

215

for learning than any other species and many of our reactions are a product of this learning, but heredity still plays an important role. How great a role is an important question. Can all people achieve equally well in all areas if given the same education or training? If so, we could concentrate on education regardless of background and end up with whatever results we desired. In certain areas, at least, this appears not to be the case. Some child prodigies have become accomplished musicians with a minimum of training, while other children may take lessons for many years with little progress. Some children seem to inherit a natural athletic ability that makes them outstanding in various events. With equal opportunities, others can accomplish little in these fields. Of course, education or training is important even for the best endowed genetically. In some communist countries, great effort is made to discover young children with natural abilities in many fields and to give them the best training so they might excel in international competitions. In this chapter we shall examine the evidence for the role of heredity in human behavior, and we shall also try to determine how environment is involved.

Inherited Human Responses

Some of our behavioral responses are almost entirely the result of heredity, others are a blend of heredity and learning, and still others seem to be almost entirely learned. The more we study human behavior, however, the more we realize that genes play a role in many reactions that we formerly attributed to learning alone. Let us see how different reactions may be classified.

Reflexes and Simple Mannerisms

Suppose you are close to the wire screen behind the home plate at a baseball game. The batter hits a foul tip that comes toward your face. You blink your eyes and duck your head, even though you know that the ball cannot possibly hit you. This is an inherited reflex, so firmly ingrained that it will occur regardless of how hard you try not to react. You have many such reflexes that are usually rather quick responses to specific stimuli that might cause harm. Other more complex reactions can also be dictated by heredity. We all have distinctive facial reactions when we recognize a friend. Some, for instance, will raise the eyebrows, others may nod the head, raise a hand, or show some other type of response. All of these responses show a strong influence of heredity, as shown by studies of siblings that have been reared apart. Even the way we place our hands when seated seems to be influenced by heredity. Identical twins, including those that have been separated shortly after birth and brought together many years later, nearly always exhibit the same type of hand placement.

Aggression

There is no doubt that people vary in their degree of aggression. Some are quickly aroused and respond strongly, while others have a very low degree of aggressive tendencies, even under threatening circumstances. Studies of this characteristic show how genes, all of which function by directing the production of polypeptide chains, can influence such complex reactions as behavior. A gene on the X chromosome results in the production of an enzyme needed for proper processing of purines. A recessive allele does not produce the enzyme, and as a consequence excessive amounts of purines and their byproduct, uric acid, accumulate in the blood. The result is **Lesch-Nyhan syndrome,** which is characterized by marked aggressive behavior. Boys who inherit this gene from their mothers will bite, kick, and often vomit over anyone who tries to touch them. They will even turn the aggression on themselves and bite their own hands, arms, and lips, often causing severe injury.

How can one gene have such an extensive effect? It appears that we all inherit nerve pathways for extreme aggression, but we learn early in life to keep such tendencies in check most of the time. Under conditions of great anger or frustration, however, the brain seems to secrete substances that stimulate the expression of aggressive tendencies. We know that the brain releases a variety of substances, known as **endorphins,** that cause certain reactions. One of these substances, **enkephalin,** acts somewhat like an anesthetic to reduce pain. Another, known as **inosine,** counteracts stress and anxiety. It is a natural product with an effect similar to the widely used antianxiety drugs such as Librium and Valium. Many of our behavioral actions may be triggered by such secretions from receptors in the brain. The products of defective purine metabolism bring out aggression in those with Lesch-Nyhan syndrome.

Gout is another inherited condition caused by defective purine metabolism. This condition is characterized by painful joints, especially the joints of the big toe, and appears more often in men than in women. The high uric acid in the blood may result in the formation of tiny crystals of sodium urate, and when these accumulate in a joint they cause great pain. Boys with Lesch-Nyhan syndrome often have gout along with their other troubles. Men with ordinary gout have normal behavior and they are often very successful. Perhaps the high uric acid level of the blood brings out just enough aggression to enable them to compete successfully in our society.

Male hormones also tend to stimulate the expression of aggressive tendencies. Boys deficient in these hormones may have difficulty adjusting to associations with their more aggressive contemporaries. Hormone injections may be necessary to correct the difficulties in social adjustment.

Another example of heightened irritability and aggressiveness is found in those who express the recessive allele for PKU (see Chapter 9). They may be irritable and have violent temper tantrums, and they typically have awkward, jerky movements because of the increase in muscle tone. The excess of phenylalanine, or some of its by-products, acts as a stimulant to bring out these latent characteristics. By controlled diet we can avoid the excess of phenylalanine and these reactions will be in the normal range.

Outside agents can also stimulate the patterns of aggression. Large doses of caffeine, which is a purine, can cause excessive aggression in rats. The nervousness that some people experience after imbibing large amounts of coffee, tea, or cola beverages may be a milder expression of this pattern. Also, the virus of rabies, when it reaches the brain, releases some substance that stimulates centers of aggression and the tendency to bite, which results in the spread of the virus to others. Some people who have developed rabies report that they have had the desire to bite those around them before they went into the paralytic stage. About 15 percent of the dogs that have rabies do not become aggressive, but have what is called the dumb type and just want to be left alone. It is possible that these dogs do not have the inherited pathways for aggression so the virus cannot bring out this characteristic.

Language

Certainly the kind of language, or languages, we speak is learned, but inheritance determines our ability to learn how to form the various speech patterns. The facility with which we learn is also inherited. Some people become multilingual with comparative ease, while others have great difficulty learning and pronouncing even a few words of a foreign tongue. Nearly all humans, however, are more proficient in learning languages during their first few years of life. A young child can often learn two languages at the same time and never get them confused when both are used in the home. This ability declines with adolescence, however, and proper pronunciation becomes more difficult. Many immigrants to the United States still have a decided accent even after living here and speaking English for many years.

The many different languages of the world did not develop by accident. Mouth and jaw structure dictates to some extent which sounds are most easily made and these anatomical features vary in different populations. The explorers who went to central Africa during past centuries found the native languages very strange and called them gibberish, but they were really very complex languages based on the mouth structure of the African people. The Neanderthals, who lived in Europe about 100,000 years ago, had a jaw structure that probably made complex speech patterns impossible. Lacking this important means of communication they were at a disadvantage when they came in competition with Cro-Magnon people who had a more modern type of jaw structure. We think that subhuman primates are not intelligent enough to learn how to speak a language, but chimpanzees have been able to master complex sign languages and some have been taught to express themselves by punching keys on a computer. They do not have the mouth structure needed for vocalizing human words. Of course, many other animals, especially birds and mammals, make sounds that convey a definite meaning, but these sounds are nothing like the complex speech patterns of humans, and instinct plays an important role in these means of communication.

Alcoholism

Most people can consume alcohol in moderate amounts without any difficulties, but for some it becomes a compelling necessity and leads to much suffering and misery for themselves and their families. Can heredity be a factor in determining who can or cannot use alcohol without becoming an alcoholic? Alcohol preference is definitely inherited in some lower animals. When mice are given a choice of pure water or water containing alcohol, some will prefer one and some the other. After several generations of selection, we can establish mouse strains where almost all prefer alcohol and other strains that are almost all teetotalers.

Human studies, however, are not so definitive. Clearly, alcoholism "runs in families," but defining the roles of heredity and environment in the determination of alcohol addiction is very difficult. We know that alcoholism may be chronic or acute, and that both forms are more common among men than among women. We know that if one's father is an alcoholic one is at higher risk for becoming addicted than if one's mother is alcoholic. The relative roles of heredity and environment can be partly disentangled by studying children who were adopted and not raised by their biological parents. In one study fifty-five Danish males, each of whom had one biological parent who was alcoholic, were adopted at an early age by nonrelatives. At the time these fifty-five men were studied, 18 percent (averaging thirty years of age) had already become addicted. Only 5 percent of the adoptees in a control population whose biological parents were not alcoholics became addicted. Another study involved the half siblings of some two hundred alcoholics. This study demonstrated that being reared by an alcoholic or nonalcoholic parent was of less significance in producing or deflecting from alcoholism than was being the biological offspring of an alcoholic or nonalcoholic. Although there clearly seems to be a biological basis for alcoholism in humans as well as in rodents, little can be said at this time about the genetic or biochemical mechanisms involved. See Ehrman and Parsons (References) for more details.

Gene Balance and Behavior

Alterations of the balance between genes can have a great influence on behavior. Trisomy-21, for instance, is one example of a chromosome aberration that causes definite patterns of behavior. Individuals having three twenty-first chromosomes have Down syndrome (Chapter 11), which is characterized by mental retardation and certain physical features, but the behavior is also distinctive. People with this syndrome are nearly always happy, contented, and have outgoing, congenial personalities. Females with Turner syndrome (XO) typically rate high on tests based on verbal comprehension, but low on tests involving perceptual organization and motor performance. Males with two Y chromosomes seem to be prone to commit crimes against property, but this is not yet fully confirmed (Chapter 5). Deletions and duplications of parts of chromosomes all play some part in influencing behavior. Each specific chromosome aberration is characterized by a particular type of behavioral alteration.

Sexual Behavior

The courtship and reproductive patterns of animals are definitely inherited and can be only slightly modified by environment. The elaborate patterns of courtship in birds, for instance, are conducted in the manner characteristic of the species even though the birds have had no previous contact with their own kind. Learning, training, and moral and ethical codes play a part in human sexual behavior, but heredity still is important. We all inherit the genes for both male and female patterns of sexual behavior, but hormonal stimulation usually makes one of these predominant while the other remains suppressed. Early training reinforces certain patterns according to sex, but heredity is still important. Although a man and woman may be physically attracted to each other, religious and moral training as well as social pressures and legal restrictions often serve as powerful deterrents to their yielding to their sexual desires.

Sometimes certain behavioral patterns will be expressed in the sex in which they are usually suppressed. **Transsexuals,** for instance, feel a gender identity that is the opposite to their chromosomal and anatomical sex. They often say that they feel like a female in a male body or vice versa. Transsexuals sometimes have sex change operations and most transsexuals say they are much happier with this altered sexual identity. **Transvestites** have a desire to dress and play the role of the opposite sex at times, although they may be normal in their other sexual reactions. **Homosexuals,** on the other hand, have a full gender identity that is the same as their anatomical sex, but they are attracted to others of their own sex as sexual partners.

The important role of environment in determining sexual identity was brought out in a study at Johns Hopkins University of six children who were 46, XX with **adrenogenital syndrome** that caused both types of sex organs to develop to some extent. Three individuals having this syndrome had been reared by the parents as females and the other three as males. Four of the six preferred the gender identity under which they had been raised. Another study in Africa by van Niekerk (1974) was done on twenty-four hermaphrodites, most of whom were known to be 46, XX; they had not been given surgical or hormonal treatment. Twenty had been reared as males and were happy with this lot in spite of the presence of female organs and breast development at puberty. The four who had been reared as females also felt full female identity in spite of the presence of penises that made them different from other females.

Another case shows how sexual behavior can be altered by hormones and also gives strong evidence for the importance of training. This case involved a pair of identical twins, one reared as a boy and the other as a girl. This strange circumstance arose because the penis of one of the boys was destroyed in an accident of an electrical apparatus that was being used for circumcision of the boys at seven months of age. The parents were advised that the twin who lost the primary male sex organ would be happier if he was given a female identity. Accordingly, they instituted surgery and hormone therapy and dressed and treated the child as a girl. She became thoroughly feminine in contrast to the twin brother who was typically masculine. The hormone therapy certainly played a part in causing expression of the genes for femaleness, but training was also important.

There are considerable differences of opinion as to the role of heredity and environment in homosexuality. Heredity gives us all the tendency to be attracted to members of both sexes, but one desire is usually predominant. Environment may reinforce latent desires for sexual contact with those of the same sex. Boys reared in homes with a strong mother and a weak father often tend to be attracted to strong, forceful men. Men and women isolated from the opposite sex, such as prison inmates, often express homosexual tendencies. In a study of forty identical twins where one was homosexual, the other was also homosexual in all cases (Kallman, 1952). More recent studies, however, have also found a high correlation for homosexuality among fraternal twins. Studies on rats show that males will show the female sexual reactions if they fail to be exposed to sufficient male hormone during early embryonic development. There is some speculation that a pregnant woman, carrying a male fetus and taking certain drugs that might suppress the early production of male hormones, could contribute to male homosexuality.

Intelligence

Intelligence lies at the base of most of our behavioral reactions; yet any attempt to ascertain the role of heredity in determining intelligence runs into many difficulties. Exactly what constitutes intelligence has not yet been completely defined or agreed upon by social scientists. Furthermore, the usual method of measuring intelligence, the IQ (intelligence quotient) test, is frequently criticized as being inadequate and biased. Intelligence, as measured by IQ tests, seems to be a multifactorial trait with both heredity and environment playing roles in its determination. These facts all contribute to the difficulties encountered in ascertaining the role of genetics in determining intelligence.

In fact, during the last fifteen years the role of heredity in determining intelligence has become a significant social and political issue in the United States and elsewhere. Some studies suggest that, on the average, black children score about 15 fewer IQ points than do white children on the standardized IQ tests. Why this is so is not clear. Some investigators have noted a 10-point daily variation in a person's performance on a given IQ test. If an individual's score can vary 10 points in one day, the 15-point average difference between blacks and whites would appear to be of much less significance. Other workers argue that the IQ test itself is culturally biased, having been developed and standardized using the white population. Consequently, black students are hampered from the outset by this test, the style and language of which are culturally foreign to them. In addition, some workers note that the U.S. black population has been culturally deprived and that the poorer performance on IQ tests of the average black is due to cultural and educational deprivation. Other investigators, notably the psychologist Arthur Jensen, have argued for the importance of heredity in determining the difference in average IQs between blacks and whites. Many researchers have been highly critical of Jensen and the statistical procedures he has used in his studies.

Many of our social and educational programs seem to be based on the premise that all human beings, other than those with definite mental defects, have an equal intellectual potential and all can be raised to the same levels if given appropriate remedial educational experiences. The recognition of the role of heredity in determining intellectual acumen is thus perceived by some individuals as a threat to such programs. Because both racial and social issues have been interjected into the discussion, obtaining a truly objective evaluation of the data is rather difficult.

Some years ago, a special investigatory committee of the Genetics Society of America, a leading organization of professional geneticists in the United States, issued a statement in the journal *Genetics* (see

References) on the race-IQ issue that was endorsed by a large segment of the organization's membership. The statement is critical of extremists advocating either a simple environmentalist or a strictly hereditarian interpretation of the data. The authors of the statement concluded that there is no convincing evidence as to whether there is or is not a significant *genetic* factor involved in causing the difference in IQ scores between the races. Some geneticists (see *Scientific American* article by Bodmer and Cavalli-Sforza) believe that obtaining definitive data on the race-IQ question is extremely unlikely because of the impossibility of conducting appropriate controlled comparative studies between black and white children raised under identical environmental conditions. The black biologist Richard Goldsby gives an interesting and nontechnical account of this controversial issue in his book, while in *The Race Bomb,* Ehrlich and Feldman attack the issue of prejudice as related to race and intelligence (see References).

Further Comments: Heredity and IQ Scores

As noted previously, IQ tests have serious limitations, but they can be useful in predicting success in academic activities. IQ scores are, no doubt, better measures of intellectual *accomplishment* than of true intellectual *potential,* but they are the best way we now have for measuring intellectual competence. Data obtained from administering IQ tests to large populations may be analyzed statistically in an attempt to determine the relative roles of heredity and environment in intellectual development. Such studies generally reveal (as one might expect) that unrelated persons who are reared apart have the maximum difference in IQs, while the IQs of monozygotic twins reared together have the highest correlation (see Table 13–1). The IQs of unrelated persons reared together exhibit a higher correlation than those of unrelated persons reared apart, thus indicating the effect of environment. Heredity, however, seems to be important when we look at the IQ scores of monozygotic twins reared apart. These scores have a higher correlation than those of unrelated persons reared together. In fact, the correlation is even higher than for dizygotic twins reared together.

In recent years, extremely interesting research on identical twins reared apart has been conducted by the psychologist Thomas Bouchard and his associates at the University of Minnesota. The studies have been concerned with both physical and behavioral similarities and differences in the twins. The research, which has been widely reported in the popular press and on television, suggests a strong genetic component in the determination of human behavior (including intelligence). An interesting and readable progress report on the Minnesota studies appeared in the November 1980 issue of *Science 80* (see References).

TABLE 13–1 Effect of heredity and environment on IQ

The degree of correlation between IQs of individuals is affected by both heredity and environment. The correlations vary from 0 (no relationship) to 1 (100 percent relationship). Having genes and/or environment in common increases the correlation between IQs. Although these data are hypothetical, they are similar to data from a number of studies reported in various journals and textbooks.

	Correlation of IQs	Genetic Correlation
unrelated children		
reared apart	0.08	0
reared together	0.18	0
dizygotic twins		
reared together	0.54	0.5
monozygotic twins		
reared apart	0.76	1
reared together	0.89	1

Ability in Different Fields

Intelligence has many facets and high ability in one area does not necessarily accompany a high level of competence in others. Some people seem to have inherited a polygenic combination that makes them virtually human calculating machines, with great facility in manipulating numbers and abstract concepts. For instance, in the early nineteenth century, Zerah Colburn, a teen-aged son of a poor Vermont farmer, could multiply any two four-digit numbers in his head and instantaneously determine the product. He required a few seconds for multiplying five-digit numbers! A woman in a mental institution could not learn how to tell time from the hands on a clock, but she could play any tune on the piano after hearing it only once. Such an individual is spoken of as an idiot savant.

By contrast, some individuals are brilliant in many fields. A boy of nine became an accomplished violinist at the Julliard School of Music, but he was also so proficient in science, mathematics, and other disciplines that he encountered serious difficulty in deciding which field should become his specialty.

Learning Ability in Rats

Some studies of learning in rats support our observations in humans that intelligence has many facets. As has been indicated, it is very difficult to measure the role of heredity in learning ability in humans because we cannot control the environments. By way of contrast, experimental animals can be reared under the same conditions so that heredity is the only variable factor. Rats, for instance, showed considerable differences in learning how to reach food through a maze. When a group of rats having mixed hereditary background, but raised under the same conditions, were faced with running the maze, some would repeatedly go up blind routes before finally reaching the food. Others learned the correct route quickly and soon went directly to the food. By breeding such "intelligent" rats with one another for eight generations, a strain was established that could learn the maze quickly. Similar models of selection for "dull" rats produced a strain of slow learners. When members of the two strains were mated, the offspring produced were intermediate in the time required for learning, an indication of polygenic inheritance.

When rats from the bright strain were given other tests, however, they did no better than average. For example, when placed in a situation where they had to dive under water to reach food, some of the "intelligent" rats did more poorly than rats from the so-called "dull" strain.

Mental Retardation

Using IQ scores as a criterion, we find a continuous range from very high to very low, with the majority of people clustered around the median (Figure 13–1). Somewhere on this scale we can draw a line and say that all who fall below the line are mentally retarded. In the past we tended to institutionalize those at the lower end of the scale; we simply gave them custodial care for their entire lives. Now we know that many such individuals can be productive members of society if given the proper training within their capabilities. Is heredity the primary cause

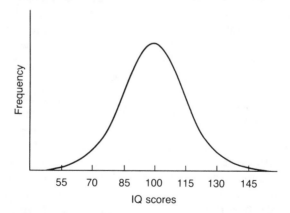

FIGURE 13–1. IQ scores of the U.S. white population are distributed in essentially a "bell-shaped" normal curve with the mean at 100. Some departures from a perfect normal distribution have been noted, however. For example, an excess of individuals with very high IQs may be attributed to selective mating. Single gene defects and chromosome aberrations may cause an excess of individuals with very low scores.

of the low level of intellectual achievement of this group, or is environment the main culprit? The answer is that heredity is the sole cause in many cases, environment is the sole cause in many others, and both heredity and environment have an effect in other cases.

Since brain structure and function are basically inherited features of the human body, we can say that heredity determines intellectual potential. Environment however, determines the extent to which this potential is realized. A zygote might receive a gene combination for a very high level of intellectual development, but if the mother is an alcoholic and drinks heavily during early pregnancy, she may have a child with severe mental retardation, the fetal alcohol syndrome, as discussed in Chapter 3. The alcohol circulating through the fetal brain would inhibit the normal development of this seat of intellect. Certain drugs, harmful chemicals, radiation, diseases such as rubella, and other agents can produce similar tragic results. Diet during pregnancy is very important. The genes may be present for normal brain development, but they must have the proper building materials. A deficiency of amino acids, vitamins, or minerals during fetal development can result in a retarded child.

The environmental effects do not stop with birth. Proper nourishment is also vital during the formative early years when the brain is still growing. Diseases, blows to the head, certain drugs, and other factors can arrest or cause regression of mental capacity. Also, mental challenge is important; the brain, like muscles, must be used if it is to develop properly. Even as adults, brain function can be retarded. When the spirochetes of syphilis get into the blood stream, some of them may invade the brain and cause paresis, commonly called softening of the brain, with irreversible loss of brain capacity. Senile dementia robs many people of their full intellectual capacity during old age. The blood supply to the brain may become deficient and some of the cells die from lack of food and oxygen. Anoxia, oxygen deficiency, can cause brain damage at any age. When breathing stops for longer than about five minutes for any reason, permanent brain damage usually results.

Certain specific genes bring about mental defects. When the gene for PKU is homozygous, the resulting symptoms include mental retardation if the diet is not kept low in phenylalanine (see Chapter 9). Single genes are also involved in such conditions as Tay-Sachs disease, galactosemia, Lesch-Nyhan disease, and cretinism; all of these conditions are characterized by mental deficiency. The nature of these diseases has been discussed in previous chapters. Huntington's disease is dominantly inherited and causes mental deterioration, but it does not usually start until well into adulthood.

In other cases, heredity confers a predisposition toward mental defect, but certain environmental conditions are necessary before the defect will appear. Schizophrenia is a good example; it will be described in Chapter 14.

Nearly all chromosome aberrations result in mental retardation. An extra chromosome, trisomy, upsets the balance of genes and one of the characteristics usually affected adversely is mental capacity. Down syndrome (trisomy-21) is the most common of these, but trisomies of longer chromosomes have an even more extreme effect on mental capacity. The sex chromosomes do not fit this pattern exactly because only one X is active in most cells; all except this one are inactivated and become sex chromatin bodies. Hence, triple-X females and XO females (Turner syndrome) are in the normal range of mental development even though one has three Xs and the other has only one X. The variation in the number of Xs, however, has some effect on mental capacity when the number is greater than one more than normal. Since one X is normal for males and two Xs are normal for females, this means that practically all males with three or more Xs and females with four or more Xs are decidedly defective mentally.

Aberrations involving the loss or gain of pieces of chromosomes also result in retardation in most cases. Even translocation, the shifting of a part of a chromosome from one place to another on the same or different chromosome can have an effect. The brain is dependent upon such a delicate balance of genes to one another that even changing their position in relation to one another can upset normal functioning. The cri-du-chat syndrome, described in Chapter 11, is one example of a chromosomal deletion that produces profound mental retardation.

Mental Retardation: Correlations in Siblings and Parents

Studies of the siblings of mildly retarded persons show that most of them are mildly retarded also, but siblings of severely retarded persons are normal in most cases (Slater and Cowie, 1971). These findings, which might seem puzzling at first, can be explained by differences in the causes of mild retardation as compared to severe retardation. The latter is usually the result of homozygosity for a recessive gene or the presence of a chromosome aberration. If a single recessive gene is responsible, the parents are most likely normal and heterozygous; and the defect would be expected in only one-fourth of their children. If a chromosome aberration is the cause, it might have appeared in a gamete of one of the parents as a result of nondisjunction, and the chance of a repeat is relatively small. Mild retardation, on the other hand, is most likely caused by accumulated polygenes, each of which contributed to a reduction of mental ability. Siblings, therefore, would be very likely to receive some of these polygenes from the parental gene pool.

When both parents are severely retarded they might both be homozygous for a recessive gene and all children would have the same retardation, but it is not likely that two severely retarded persons would marry. If only one parent is severely retarded, as in cases of rape of a severely retarded woman, the chance of a retarded child is small because the normal parent is not likely to be a carrier of the recessive gene. A chromosome aberration, for example a 14/21 chromosome translocation, might be transmitted in the same manner as a dominant gene. On the other hand, mildly retarded parents will be likely to pass some of the polygenes involved to their children. Of course, all evaluations must consider the possibility that the retardation, either mild or severe, is caused by environmental factors and would not be transmitted genetically.

Social Policies and Mental Retardation

The recognition that intelligence and mental retardation are affected by heredity as well as environment led to the following questions: Will mentally handicapped parents have children like themselves? If so, does society have a right to limit their reproduction, since their offspring are likely to be a burden on society?

Those who supported the **eugenics movement** in the first four decades of this century answered both of these questions with a resounding "Yes!" In 1907 the state of Indiana passed the first sterilization law. When approved by a board of experts, the law mandated the sterilization of imbeciles, idiots, criminals, and others in state institutions. Although Indiana had the dubious distinction of being the first, it certainly was not the only state to pass a sterilization law. By the early 1930s such laws had been enacted in over 60 percent of the states in the United States.

The supporters of eugenic sterilization also found a receptive audience among the leadership of Nazi Germany. Sterilization became a significant tool in the Nazi's misguided quest for racial purity and superiority. Not only were the mentally retarded sterilized, but also many others whose political philosophies were unacceptable to the ruling Nazis. The eugenics movement, which had generally fallen into hands of racists and crackpots, lost its last vestige of respectability with the many excesses committed by the Nazi regime.

The question still remains, however: Would sterilization of the mentally defective significantly improve the human gene pool? The evidence is largely negative. Most mentally retarded individuals are produced, not by the mentally retarded, but by those of us who happen to be heterozygous carriers of the recessive genes that cause retardation when homozygous. If, for example, we wanted to eliminate PKU (which can cause severe mental retardation), we would have to sterilize not only the 1 in 10,000 individuals who is a PKU victim, but also the approximately 1 in 50 who is a "normal" heterozygous carrier of the recessive gene for PKU. Obviously, this "solution" to one cause of mental retardation is impractical and would be totally unacceptable to society.

Involuntary sterilization to prevent mental retardation has largely been abandoned. The chance for abuse of such a program is too great, and the genetic worth of the program is highly questionable.

PROBLEMS AND QUESTIONS

1 As a rule, the groups of animals that have a very short life cycle also have behavioral reactions that are primarily instinctive, while learning can play an important role in those animals with long life cycles. Give a possible explanation for these differences.

2 Describe a human pattern of behavior that seems to be due almost entirely to heredity. Justify your answer.

3 Describe a human behavioral characteristic that seems to be the result of environmental influences with little hereditary influence. Justify your answer.

4 Identify a human pattern of behavior where both heredity and environment seem to be involved, and tell how you would use the results of twin studies to assess the relative role of each of these forces.

5 Aggression involves a number of complex behavioral patterns, yet it can be brought out in a strong degree of expression by a mutation of a single gene. Explain how this can be possible.

6 If human babies could be raised completely isolated from adults, do you think they would develop a language of their own? Explain.

7 Mental retardation characterizes nearly all people who have chromosome aberrations in their cells, even though no genes may be missing. Explain.

8 How have studies of hermaphrodites shown the importance of training in developing patterns of sexual behavior?

9 Consider again the isolated babies described in Question 6. Would you expect as much homosexuality among them as they matured as we have in our society? Justify your answer.

10 In many large families one child will have a severe mental defect, but the other children are normal. In many other large families, however, when one child is mildly mentally retarded there is a high probability that the other children will also be mentally retarded. Does this mean that heredity influences mild retardation, but not severe retardation? Explain.

11 Schizophrenia is brought on by extreme stress, yet we know that heredity is involved. Explain the relationship.

12 Explain the overall role of heredity and environment in the establishment of intelligence.

REFERENCES

BODMER, W. F., and L. L. CAVALLI-SFORZA. 1970. Intelligence and race. *Scientific American* 223(4):19–29.

EHRLICH, P. R., AND S. S. FELDMAN. 1977. *The Race Bomb. Skin Color, Prejudice, and Intelligence.* Quadrangle/The New York Times Book Co. New York.

EHRMAN, L., and P. A. PARSONS. 1981. *Behavior Genetics and Evolution.* McGraw-Hill Book Co. New York.

FULLER, J. L., and W. R. THOMPSON. 1978. *Foundations of Behavior Genetics.* The C. V. Mosby Co. St. Louis.

GOLDSBY, R. A. 1977. *Race and Races,* 2d ed. Macmillan Publishing Co., Inc. New York.

GOTTESMAN, I. I., and J. SHIELDS. 1972. *Schizophrenia and Genetics: A Twin Study Vantage Point.* Academic Press. New York.

HOLDEN, C. 1980. Twins reunited: More than the faces are familiar. *Science 80* 1(7):55–59.

LEWONTIN, R. C. 1975. Genetic aspects of intelligence. *Annual Review of Genetics* 9:387–405.

MITTWOCH, U. 1973. *Genetics of Sex Differentiation.* Academic Press. New York.

MORTON, N. E. 1972. *Genetics, Environment, and Behavior.* Academic Press. New York.

PLOMIN, R., J. C. DeFRIES, and G. E. McCLEARN. 1980. *Behavioral Genetics: A Primer.* W. H. Freeman. San Francisco.

RUSSELL, E. S., et al. 1976. Report of the *ad hoc* committee: Resolution on genetics, race and intelligence. *Genetics* 83(3–1):s99–s101.

SLATER, E., and V. COWIE. 1971. *The Genetics of Mental Disorder.* Oxford University Press. London.

fourteen

Heredity and Environment

WHEN WE FOCUS OUR ATTENTION ON THE powerful effect of genes, as we have in this book, we tend to overlook or underestimate the important role of environment in the total development of an individual. Some traits, such as the blood types, are determined by heredity and are not significantly influenced by environment. Other traits seem to be determined primarily by the environment: a child may be born deaf because of a rubella infection of the mother during the pregnancy; or a child may develop a severe case of rickets because of a diet deficient in vitamin D. Both deafness and rickets can also be produced by genes, however. Most traits reflect an interplay of both nature (heredity) and nurture (environment). One's stature, for example, is under the control of polygenes (see Chapter 4), but it also can be markedly affected by diet and communicable disease. Likewise, intelligence (Chapter 13) appears to be a polygenic trait that is markedly influenced by environmental factors. One fact seems clear: before we can assign a genetic or an environmental cause of any trait, we must investigate carefully.

We can compare the roles of heredity and environment to the exposure and development of a photographic film. You load your camera with color film and make some exposures, yet you have no image on the film. There is a potential image, however, which can be brought out by the appropriate developing agents. If you make a good exposure and the proper chemicals are used for the proper time and at the proper temperature in development, you will have a good color photograph. Even with the best exposure, however, poor development will not realize the potential. Likewise, if the exposure is poor, then no amount of good development can make a good photograph, although it can realize the most possible for the potential. Heredity gives the potential and envi-

231

ronment functions in a manner analogous to development. In this chapter we shall see how these two important forces interact.

Environmental Modification

Environment sometimes profoundly alters the influence of genes. For instance, the degree of skin pigmentation varies greatly according to one's genetic background, yet exposure to sunlight can also greatly alter the degree of melanin deposited. A Hawaiian boy who has inherited genes for a medium amount of melanin deposit will have a nice bronze skin regardless of whether he ever gets in the sun. A fair-skinned Norwegian boy living in Hawaii may spend so much time on the beach that he becomes as dark as the native Hawaiians.

Phenocopies

Environment sometimes can produce a condition that is an exact duplicate of a trait that is genetic at other times. This is termed **phenocopy.** For example, a recessive gene causes **phocomelia** (partial absence of arms and legs as described in Chapter 3). This condition appeared at rare intervals for centuries because the gene became homozygous (Figure 14–1). In Europe in 1961 and 1962, however, there was an alarming increase of phocomelia because pregnant women had been taking the drug thalidomide. This drug slowed the development of arms and legs at a critical time, resulting in a phenocopy of genetic phocomelia.

Hydrocephalus, or water on the brain, is characterized by a greatly swollen cranium. Cavities in the brain, called ventricles, contain the cerebrospinal fluid, which must be maintained at a certain pressure to hold the soft brain out against the bones of the cranium. Pressure is regulated by absorption of some of the fluid by the membranes around the ventricles and by drainage through a duct into the cavities around the spinal column. An autosomal recessive gene causes a thickening of the membranes, which slows absorption. The result is a buildup of fluid in the ventricles, and this causes the brain to swell. The bones of the cranium are soft in the developing fetus and the internal pressure causes them to expand, resulting in a baby with a very large cranial region of the skull. Another gene, an X-linked recessive, causes an obstruction of the drainage duct and has a similar effect. A phenocopy of hydrocephalus can also result when the mother has a virus infection during early pregnancy when the membranes around the ventricles are forming. Heavy radiation given at this critical time will also cause the same condition. Hence, if someone asks if hydrocephalus is inherited, no definite answer can be given. It sometimes can be the result of heredity; other times environment is the cause. Only pedigree analysis

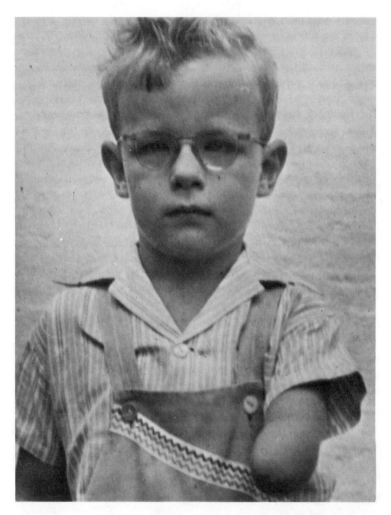

FIGURE 14–1. Phocomelia, partial absence of a limb, was caused in this boy by a recessive gene received from each parent. Many cases in Europe during the early 1960s were induced by thalidomide taken in early pregnancy by the mothers. Thalidomide-induced cases of phocomelia are thus phenocopies and are not heritable.

and a study of the experiences of the mother during pregnancy can indicate which might be the cause in any particular case.

We can produce phenocopies of many genetic defects in experimental animals. Monkeys can be induced to show the symptoms of PKU when given much greater than normal amounts of phenylalanine in their diets. Thus, a phenocopy of PKU can be produced by environmental modification. We can also produce a phenocopy of galactosemia by feeding rats diets very rich in galactose.

Variable Expressivity

People with the same genotype for certain genes can show a great variation in the degree of expression of these genes. Those who are homozygous for the gene for PKU and who take no treatment show variation in mental capacity from the low idiot level up to the moron level. Variation in the intake of foods containing phenylalanine could account for some of this variation. When the intake of phenylalanine is purposefully restricted, the persons may have normal mentality.

Other genes within the cells may account for many cases of variable expressivity. Figure 14–2 shows a father and his son. Both have flipper arms, the result of the action of a dominant gene, but the reduction in arm size is much more extreme in the son than in the father. Since no external environmental factor seems to be involved, it is probable that differences in other genes have caused this variation in development.

The expressivity of a gene may vary among the different characteristics of a syndrome. **Osteogenesis imperfecta** is a syndrome that usu-

FIGURE 14–2. Variable expressivity of the dominant gene for flipper arms. The father, at right, has a better arm development than his son, at left. Modifying genes or prenatal environment can alter the expressivity of genes such as this one. (Courtesy Karl Stiles.)

ally results from a dominant gene. This gene causes improper phosphate metabolism; an excess of phosphate is found in the urine, but there is a deficiency in the body cells. About 90 percent of those individuals having the gene have blue sclerae. The sclera, or "white" of the eye, is blue and the density of the color can range from very light to very dark. About 70 percent of the people having this gene have brittle bones to varying degrees. The bones may be so brittle that one can break a leg by catching a toenail on the sheet while turning over in bed. In others the bones can stand much more strain, although they break more easily than normal bones. About 50 percent of the people having this gene develop deafness, usually beginning in the teens, because of a hardening of the ear bones. Yellow teeth and a thin, transparent skin are other characteristics that may be present in varying degrees.

Reduced Penetrance

Early studies of human heredity were sometimes confused when a trait that appeared to be dominantly inherited was expressed in a child when neither parent had it. We now know that such cases can be explained by **reduced penetrance.** The dominant gene for **polydactyly** is an example. Most people who have this gene have six fingers on each hand and six toes on each foot (see Figure 14–3), but there is considerable variation in expression of polydactyly. Occasionally a person possesses the gene and has the normal five digits on both hands and feet. Other genes or certain environmental factors during intrauterine life may prevent the expression of genes with reduced penetrance. About 5 percent of the individuals who have the dominant gene for Huntington's disease never show the symptoms; therefore, Huntington's disease can be said to have 95 percent penetrance. The age of onset of this disease varies considerably and it could be that the 5 percent who escape have the expression delayed so long that they die of other causes—heart disease, for example. The gene for osteogenesis imperfecta has a 100 percent penetrance in the part of the syndrome for abnormal phosphate metabolism, but the penetrance varies from about 50 percent to 90 percent for other phases of the syndrome.

Recessive genes can also have reduced penetrance when homozygous, but these cases are more difficult to recognize because it is hard to know if individuals are homozygous when they do not express the trait.

We now use various techniques to alter the penetrance of harmful genes. The recessive gene for galactosemia formerly had a 100 percent penetrance, but now that we have learned to withhold milk from the diet as soon as symptoms appear, the penetrance has been reduced to a very low level (see Chapter 9). The gene is expressed only in those homozygotes who do not go on the diet.

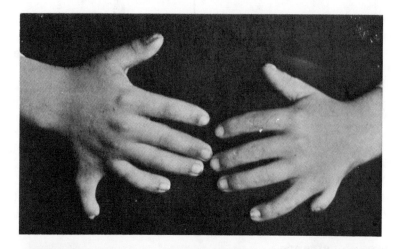

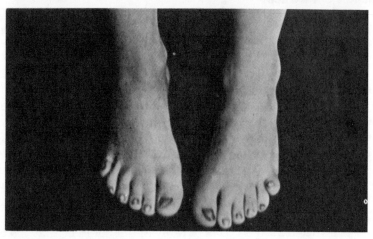

FIGURE 14-3. Polydactyly, the presence of extra fingers and toes, results from a dominant gene with variable expressivity and incomplete penetrance.

Schizophrenia

More than half of the patients in mental institutions have some form of schizophrenia. Schizophrenia is characterized by a confusion of reality with fantasy. Some express it as paranoia, a persecution complex. Others show a catatonic withdrawal and are oblivious to their surroundings. Some have delusions of grandeur, while still others have what we would call silliness. Schizophrenia may appear at any age but is most common in young adults and typically expresses following a period of great emotional stress. This latter fact led some psychiatrists to conclude that it was environmentally induced and that heredity was not involved. Continuing study, however, has shown that familial background is an important factor. The confusion arose because of reduced penetrance of genes that may be involved. Some people seem to inherit the genetic potential but never come under sufficent stress to develop schizophrenia.

Serotonin is a secretion of brain cells and is necessary for normal brain function. There may be an oversecretion of this substance in individuals with schizophrenia. Monkeys can be made to show schizophrenic symptoms when serotonin is injected into their brains. Secretions of the adrenal glands are also involved. About 1 or 2 percent of our population seems to inherit genes for abnormal secretions of these substances when under great stress. Some persons with this genetic potential may go through life without showing any serious manifestations of the abnormality, although some may have schizoid personalities.

Variability in the degree of expression of schizophrenia has led some geneticists to a theory of polygenic inheritance. Other scientists have suggested possible dominant inheritance with reduced penetrance of the gene. There is also good evidence for a single gene with variable expressivity and an intermediate effect. When homozygous for this gene, a person is very likely to react to stress by an upset in the balance of secretions, and some form of schizophrenia results. Heterozygous persons may also react in this manner, but a greater amount of stress may be required. Some feel that even people homozygous for the normal allele might become schizophrenic by excessive and prolonged stress. Figure 14–4 shows these relationships.

Twin Studies

Many of the questions related to the influence of heredity and environment can be answered by comparisons of the two types of twins. Since identical twins have the same genotype, any differences between them must be due to environment (Figure 14–5). Fraternal twins, on the other hand, have many genetic differences, so both heredity and environment can be involved in determining phenotypic differences. Because all identicals are of the same sex and sex differences in fraternals could introduce a significant bias in conclusions, we must consider only fraternal twins of the same sex. Particularly valuable for such studies are identical twins reared apart. In some cases, circumstances result in the twins being separated, with one being adopted by one family, and the other being taken by a different family. Raised by different parents, perhaps in different parts of the country, these twins will show the maximum influence of different environments on the same genotype.

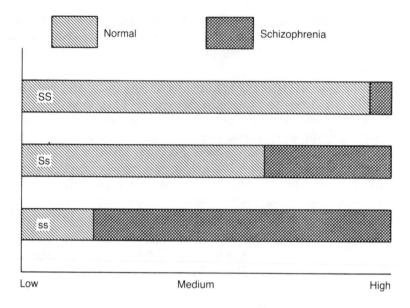

FIGURE 14–4. Proposed relation between heredity and environment in schizophrenia. According to the theory depicted here, a person homozygous for a certain intermediate gene requires comparatively little stress to develop the mental illness. A heterozygous person requires a greater degree of stress. Extraordinary stress may even bring on the symptoms in those not having the intermediate allele. Other gene loci may also be involved.

We can also consider same-sexed brothers or sisters born at different times provided we make allowances for the differences in age. These individuals are known as **sibs,** a shortened form of the term **siblings.** Their gene differences would be the same as those for fraternal twins.

Concordance

The degree of similarity between twin pairs with respect to a particular trait can be measured by **concordance,** which is usually expressed in percentages. A high percentage of concordance among identical twins and a much lower percentage of concordance among fraternals would indicate a strong influence of heredity. A concordance that is about the same for both types of twins would indicate a predominant influence of environment.

The concordance for **albinism** in identical twins is practically 100 percent. This means that when one twin is an albino, it is almost certain that the other twin also will be an albino. (We say *almost* because one case of discordance has been reported, probably the result of a

FIGURE 14–5. Environmental alteration of genetic potential for stature in identical twins. The twin on the left had a severe infection at five years which retarded development of the part of the pituitary gland that produces growth hormone. (Courtesy Taku Komai and G. Fukuoka.)

mutation in one shortly after it split off from the other.) In fraternal twins, however, the concordance is only about 25 percent, as expected, since the chance of albinism from heterozygous parents is one-fourth. These results show that albinism results from heredity and that the ratio is not altered by environment.

Harelip is the result of incomplete fusion of parts of the upper lip during embryonic development. The concordance for identical twins in one study was 33 percent and only 5 percent for fraternals. Such a low percentage for identicals shows a strong environmental influence that can affect one twin and not the other, even though both are in the same uterus. The difference in concordance between the two types of twins, however, also indicates an influence of heredity. Certain drugs and hormones have been found to induce harelip in mice, but these drugs and hormones are much more effective in certain genetic strains than in others. Evidently, a certain genetic predisposition makes the harelip more likely to occur.

We have learned that **deafness** often results in children of mothers who have rubella in early pregnancy. For such deafness, concordances of 86 percent and 88 percent were found for identical and fraternal twins, respectively. Such a close agreement shows that this trait is due to environment. That there are discordances of 14 and 12 percent indicates that there can be a small difference in the time of ear development or degree of infection in a pair of twins.

Table 14–1 shows the concordance percentages for a number of traits. Note that even susceptibility to infectious diseases can be influenced by heredity. Differences in the inherited immunoglobulins can be involved (see Chapter 7).

TABLE 14–1 Twin concordance (percent similarity) for certain characteristics

Trait	Monozygotic Twins	Dizygotic Twins
measles	95	87
tuberculosis	65	25
diabetes mellitus	84	37
rickets	88	22
cancer	61	44
site of cancer (when both have cancer)	95	58
clubfoot	32	3
harelip	33	5
Down syndrome (mongolism)	89	6
epilepsy	72	15
schizophrenia	86	15

Rickets is a noninfectious disease that is caused by a deficiency of vitamin D, yet Table 14–1 shows that heredity plays a part. Genes determine the level of vitamin D necessary to be protected from the disease. One gene causes vitamin-D resistant rickets. Massive quantities of vitamin D are required to prevent the expression of this gene.

Intelligence shows continuous variation, but at some point on the scale we say that those below a certain IQ are mentally retarded. Twin concordance shows that although heredity certainly plays a part, genes can only give a potential. Without proper food for brain development before and in the early years after birth, the potential cannot be realized. Also, improper training of the intellect can decrease the amount of potential that is realized. For these reasons, some persons might be classified as mental retardates when they actually have inherited genes for normal mental development.

Quantitative Differences

Traits that differ quantitatively are usually measured by differences in the two types of twins. Table 14–2 includes a number of traits of this kind. The figures indicate that body stature is strongly influenced by heredity and body weight also, but to a lesser degree. Differences between monozygotic twins reared apart and those reared together show the influence of environment on IQ.

TABLE 14–2 Twin and sib differences in quantitative traits

Trait	Monozygotic Twins	Dizygotic Twins	Sibs	Monozygotic Twins Reared Apart
stature (diff. in cm)	1.7	4.4	4.5	1.8
weight (diff. in lbs)	4.1	10.0	10.4	9.9
age of first menstruation (diff. in months)	2.8	12.0	12.9	—
IQ (Binet) (diff. in points)	5.9	9.9	9.8	8.2

Heredity and Environment in Cancer

Cancer is one of society's most important medical problems. Almost everyone has had close relatives or friends who have suffered and perhaps died as a result of the uncontrolled growth of tissue known as cancer. It is the direct cause of about one death in each five in the United States and currently one in every four persons can be expected to develop cancer. Your chance of having cancer, therefore, is quite high, but the more you know about it the greater is the probability that you can escape its destructive actions. You may be confused by the many stories about the cause of cancer and many questions may have arisen in your mind. Is it caused by heredity, environment, chance, or some combination of these? We can better understand the possible causes if we know more about the nature of cancer and its variations.

The Nature of Cancer

During fetal life, cell growth, division, and differentiation occur in all parts of the body as organs are formed and increase in size. These processes continue after birth until adult size is reached, at which time the rapid growth stops, although some types of cells continue reproduction at a rate that replaces those that are worn out and destroyed. For instance, cells of your outer skin and the lining of your intestine are constantly flaking off and being replaced by cell divisions of the remaining cells. If you injure your skin, some force stimulates increased cell division until the damaged tissue is restored, and then cell growth and division reverts back to the replacement level. A precise regulation is involved and it sometimes fails, resulting in unregulated growth. Cells grow wild and form undifferentiated tissue that is not a normal part of the body. The tissue becomes a parasite, deriving nourishment from the body but contributing nothing to it. Such abnormal growth is known as a **tumor,** and it may be benign, which means that it does not spread to other parts of the body, or it may be malignant and can spread.

Terminology about cancer is often confusing, so let us be certain we understand necessary terms before going further. Any uncontrolled growth is a **neoplasm** or **tumor. Malignant neoplasms** can spread by growth into adjoining tissues, and some of the cells can break loose and be carried by the lymph and blood to other parts of the body where they can start new growths. Such malignant neoplasms are commonly known as **cancers.** They may be **carcinomas,** neoplasms of epithelial tissues such as are found in the skin or the linings of the lungs and intestine, or **sarconomas,** neoplasms of connective tissue, which is a part of many organs of the body. Sometimes neoplasms are referred to according to the part of the body affected. A **melanoma** is a neoplasm of pigmented cells of the skin, an **adenoma** is a neoplasm of glands, a **myoma** is a neoplasm of muscle tissues, and **leukemia** is a neoplasm of the cells in the bone marrow that produce leukocytes. Any agent that induces a cancer is a **carcinogenic agent** or an **oncogenic agent.**

Causes of Cancer

Cancer may be induced in a number of ways that superficially seem to be unrelated, although two or more causative agents may be involved in some cases. Many environmental carcinogens—both natural and man-made—may produce their effects by damaging (mutating) cellular DNA. Damage to the DNA of germ cells can cause genetic defects that will appear in future generations. Mutation of DNA in somatic cells of the body may give rise directly to cancerous cells by altering the normal cellular mechanisms that control or prevent cell multiplication.

Genes   The suggestion that genes play a role in the etiology of cancer is not a new idea. In 1914 the German zoologist Boveri set forth the "somatic mutation" hypothesis for the origin of cancer. Boveri's hypothesis suggests that at least some cancers result from somatic mutations that alter growth patterns of cells in the affected body part of the cancer victim. As somatic mutations, such genetic changes are not transmissible. In addition, single genes, some dominant and some recessive, can bring about conditions conducive to cancer. Several examples of such genes will be discussed below. Also, polygenes with additive effects seem to be responsible for some cases of cancer. Such single gene and polygene conditions can be transmitted from generation to generation.

Multiple polyposis (Gardner syndrome) is inherited through a dominant gene. This gene causes the growth of many mushroomlike polyps on the interior of the large intestine. If not removed surgically, some of these polyps will very likely become malignant neoplasms. While the polyps may develop during childhood, they usually do not become malignant until after sexual maturity, so the gene can be passed to children by unknowing parents.

Retinoblastoma is a malignant growth in the retina of the eye. Although characteristically caused by an autosomal dominant gene, some cases of familial retinoblastoma have been clearly demonstrated to be associated with a deletion in the long arm of chromosome 13 (i.e., $13q^-$). Retinoblastoma leads to blindness and eventual death unless the eye is removed or the malignancy destroyed by radiation before it spreads. About 1 child in 25,000 develops retinoblastoma; the disease accounts for about 1 percent of all cancer deaths in infancy. Only about 15 percent of the cases are passed from an afflicted parent to children; the other 85 percent seem to arise as somatic mutations in cells of the retina. When both eyes are affected it is probable that the gene came from a parent, but when only one eye is affected it is usually an indication that a somatic mutation is the cause. A number of cases have been traced to excessive exposure of the eyes to ionizing radiations.

Xeroderma pigmentosum is an example of a recessive characteristic that can lead to cancer. A baby homozygous for the gene will be normal at birth, but the skin is extremely sensitive to ultraviolet light, and even brief exposure to sunlight will cause large, dark freckles to appear. Some of these become hard, warty lumps that may become malignant. Cancer claims the lives of most people with this affliction before the age of thirty. Apparently, they cannot produce the enzyme **endonuclease,** which repairs broken DNA chains. Ultraviolet light causes such breaks in all people, but the repair usually is made in those with the enzyme. Unrepaired chains can lead to malignant growth.

Some other recessively inherited diseases increase the risk of cancer, possibly because they increase the number of chromosome aberrations, and certain chromosome aberrations are known to lead to cancer. The cause again seems to be an enzyme deficiency, in this case the enzyme that can heal chromosome breaks. **Bloom syndrome** is one such disease. Individuals homozygous for the gene are characterized by retardation of growth and dilation of capillaries in the skin that gives many red, spiderweblike patterns on the skin (telangiectasia), especially on the cheeks. The term butterfly rash is sometimes used to characterize the pattern on the cheeks. Karyotypes made from people with the disease show many chromosomal abnormalities. In addition, blood analysis shows a decrease in immunoglobulins IgA and IgM. Since immunoglobulins help overcome cancer, this may be a factor in the high incidence of cancer in those with Bloom syndrome.

Fanconi's anemia is another recessively inherited disease that increases the chance of chromosome aberrations and cancer. It is characterized by a defect of the cells in the bone marrow that produce blood cells and by major abnormalities of the skeleton, heart, and kidneys. Leukemia is a type of cancer that is most commonly associated with this disease.

Polygenes seem to be involved when cancer runs in families but does not show typical dominant or recessive inheritance. A woman's chance of having breast cancer, for instance, is increased if her mother, sister, or other close relative has had it. If the mother had the cancer before menopause, then the daughter's chance is about 6.7 percent as compared to 2.3 percent in controls where the mother did not have breast cancer. If the mother's cancer involved both breasts and was premenopausal, the chance for the daughter jumps to 17 percent. When more relatives are involved the chance goes up. Two sisters, the mother, the grandmother, and several aunts and cousins of one young woman had undergone surgery to remove breast tissue because of cancer. A geneticist-physician calculated the woman's chance of having breast cancer at 50 percent. Twin studies indicate a concordance of about 28 percent for monozygotic twin women and 12 percent for dizygotic women. These results indicate that several gene loci are involved in susceptibility to breast cancer; some environmental factors could also be involved. The genes can be inherited from both parents, but are rarely expressed in normal males. Males with Klinefelter syndrome, however, express the genes about as frequently as females.

The frequency of skin cancer is correlated with the degree of melanin deposit. Blacks seldom have skin cancer, but the frequency is high in fair-skinned whites, depending on the degree of exposure to sunlight (Figure 14–6). Since we know that polygenes are involved in determining the amount of melanin in the skin, this would appear to be another case where polygenes with an intermediate expression influence the chance of cancer induction by an environmental agent.

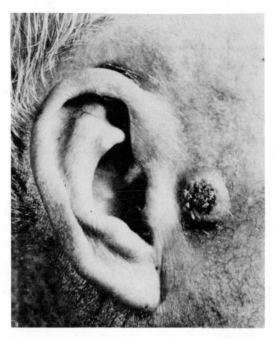

FIGURE 14–6. A skin cancer. Members of the white races are especially susceptible to this type of cancer.

Lung cancer is also a case where both hereditary and environmental factors are involved. People vary in the degree of production of an enzyme (aryl hydrocarbon hydroxylase) as a result of their genetic background. This enzyme acts on the hydrocarbons found in cigarette smoke to convert them into an epoxide that is carcinogenic. A measurement of the amount of this enzyme produced by a randomly selected sample of people showed that about 45 percent had a low concentration, 46 percent had an intermediate concentration, and 9 percent had a high concentration. Such results indicate that a single pair of alleles are involved and that they have an intermediate influence when heterozygous. An analysis of the amount of this enzyme in smokers who had developed carcinoma of the lungs showed almost none with a low level, about 70 percent with an intermediate level, and about 30 percent with a high level. Table 14–3 shows this relationship. Note that the risk of lung cancer goes up in those smokers who have the higher enzyme levels.

Chromosome Aberrations   Karyotypes made from cancer cells frequently show chromosome aberrations. Translocations, deletions, duplications, aneuploidy, and polyploidy are all found, but it is not yet clear whether the aberrations appear first and cause the cancer, or whether the cancer appears first and causes the aberrations. The type

TABLE 14–3 Relation between aryl hydrocarbon hydroxylase concentration and lung cancer in smokers

Concentration of Enzyme	Percentage in Each Group	Percentage with Lung Cancer	Risk Factor $\dfrac{\% \text{ with Cancer}}{\% \text{ of Smokers}}$
low	45	Near 0	Near 0
medium	46	70	1.52
high	9	30	3.33

Data from Fraser and Nora.

of aberration may be different in persons with the same type of cancer so there generally does not seem to be any association of a particular type of aberration with a certain type of cancer. One important exception, however, seems to occur in most cases of chronic myeloid leukemia (CML), where bone marrow cells of affected individuals usually possess a translocation of a part of the long arm of chromosome 22 to chromosome 9, as was described in Chapter 11. Another case does not involve a particular chromosome, but is associated with a type of aberration. People with Bloom syndrome have a high frequency of sister chromatid exchanges.

Persons with syndromes resulting from chromosome aberrations are more susceptible to cancer in some cases. Individuals with Down syndrome, for instance, have about 1 chance in 96 of developing leukemia, while the frequency is only 1 in 2880 for the general population, a thirty-fold difference. Other trisomies, such as Klinefelter syndrome, also are associated with an increased risk of cancer.

Oncogenes and Viruses   Some 20 oncogenes cause cancers in various animals and induce cultured animal cells to undergo malignant transformations. To date, only circumstantial evidence implicates oncogenes in the production of human cancers, but evidence is mounting that such cancers have their origin in the activation of oncogenes. The transformation of a normal, healthy cell to a malignant condition may result from an increased production by a "normal" gene of a protein product, such as a growth factor. In some cases the transformation may be caused by the manufacture of a defective gene product. The production of the malignancy may even require the concurrent activation of more than one oncogene.

Any discussion of oncogenes requires consideration of the role of viruses in causing cancer. As early as 1911, F. P. Rous demonstrated the role of a virus in causing sarcomas in chickens. Subsequently, other viruses were shown to play a role in tumor formation. Most of these cancer-inducing viruses have RNA as their genetic material; they are called **retroviruses** because they produce the enzyme reverse transcriptase that copies RNA into DNA (the reverse of the "normal"

DNA to RNA sequence occurring in protein synthesis). The ability of the Rous sarcoma virus to cause cancer has been attributed to the protein product of a specific gene, designated *src*, in the virus's genetic makeup, but the exact mechanism of transformation is not yet clear. Normal cells of the chicken contain DNA sequences that are similar to the *src* gene of the Rous virus. The *src* gene also has been located in human chromosome 20, although no tumor has been associated with it in humans.

Although oncogenes were first identified as the genetic factors of the retroviruses responsible for malignant transformation of host cells, it is now clear that they are actually the genes of higher organisms. At least 16 of the 23 human chromosomes have been shown to possess loci for such oncogenes. Some evidence suggests that human oncogenes play a role in initiating tumors in the absence of viruses; e.g., certain normal mouse cell lines, treated with DNA from human tumors, undergo malignant transformations. Specifically, variants of the *ras* oncogene isolated from human bladder carcinoma cells can cause the transformation of mouse cells to the malignant state. Second, certain malignancies are associated with chromosome rearrangements and only tumor cells of the patient possess the rearrangement. Studies of tumors associated with chromosome aberrations have led to the hypothesis that the aberration must modify an oncogene to a form capable of producing a malignancy and at the same time stimulate it to express at a high level. This appears to be the case with the Philadelphia chromosome associated with chronic myelogenous leukemia in humans. The chromosome aberration in this case is a reciprocal translocation between chromosomes 9 and 22 (see p. 190). In **Burkitt lymphoma** the *myc* oncogene seems to be rearranged as a result of a translocation involving chromosomes 8 and 2, 14, or 22. Following translocation, *myc* is transcribed to RNA at elevated levels as compared to normal cells.

Additional research has shown that other retroviruses have oncogenes that seem to have been acquired from the DNA sequences of normal host cells. Viral oncogenes seem to be erring copies of genes found in a variety of animal species, including mice and humans. The cellular forms of these genes often seem to have major roles in cell differentiation or regulation of cell division, but as oncogenes they misbehave and produce the transformation to malignancy.

Carcinogenic Agents   Back in the 1930s it was found that where certain coal tar derivatives were applied to the skin of mice a neoplastic tumor developed. Since then, many other substances have been found to be carcinogenic. It is often difficult, however, to translate findings in experimental animals to humans. Quantities used in experiments are usually much greater proportionately than those to which people would be exposed. As artificial sweeteners, cyclamates have been banned and saccharin has been condemned because they were found to be cancer-causing when fed to experimental animals in large quantities. Some have argued, however, that the much smaller quantities used in human foods were not enough to be of significant danger. The same is true of red dye number two, which was banned after being used for many years to color preserved cherries, candies, soft drinks, and so forth. Sodium nitrate is still used extensively in small quantities to preserve ham, bacon, processed meats, and canned meats. Some people inherit an enzyme that converts it into nitrosamides, which have been shown to be powerful carcinogens in higher concentrations in experimental animals. We have already mentioned a similar situation for the hydrocarbons of tobacco smoke. Some pesticides and herbicides have been banned because they have been found to be carcinogenic. In one year the entire cranberry crop was kept off the market because residues of such a pesticide were found. Certain gases used in manufacturing processes have been alleged to cause cancer of the lungs, liver, and kidneys of workers in plants where they were used. These have now been banned. A flame retardant added to children's nightwear was found to be carcinogenic and has been taken off the market.

Radiation treatment has been a mixed blessing. During the latter part of the nineteenth century, Marie Curie found that ionizing radiation given off from radium, if administered in just the right amount, would destroy cancer cells without apparent harm to the normal surrounding tissue. Ionizing radiation is still one of the most effective weapons against neoplastic growths. It is one of the ironies of fate, however, that such radiation can also induce cancer in healthy tissue. Many cases have been found where cancers developed at the site of radiation exposure, sometimes many years later. Thyroid cancers have developed in about 7 percent of the persons who had radiation treatment of the neck region for reduction of a large thymus gland or of the face for the treatment for acne. The treatments were given in childhood, but the cancers were not evident for as long as twenty years in some cases. Many cases of cancer were noted in those exposed to the radiation of the atomic bombs in Japan, often many years after the exposure. The mobile chest X-ray units, used for many years for mass screening to uncover incipient cases of tuberculosis, have been discontinued because evidence indicated that the danger of cancer induction and possible genetic damage outweighed the good done. Besides, other methods can be used for the screening. Also, mammographs are widely used to detect early breast cancer, but some feel that the amount of X rays used could pose a carcinogenic threat.

An important question is involved in making decisions on the medical and economic use of ionizing radiation. Is there a threshold below which no cancers are induced? This is indeed a very important question in the light of the many very small exposures received in routine physical checkups. Also, we must consider the possibility of genetic damage, as was brought out in Chapter 10. Although we do not have a conclusive answer to this question yet, we do know that the risk of cancer induction is proportional to the dosage in the higher exposures.

An overview of the various factors that may be involved in cancer induction brings out some important correlations. Agents that are mutagenic (cause both point mutations and chromosome aberrations), teratogenic (cause abnormalities in the fetus when pregnant women are exposed), or immunosuppressants are very likely to be carcinogenic as well. Ionizing radiation, for instance, has all these effects, but some agents can cause one type of damage without the others. Any agent that causes one effect, however, should be highly suspect for the others because the agents are all closely related. A chromosome aberration in a gamete can result in a defective child, in a somatic cell of an early embryo it can be teratogenic, and in a somatic cell after birth it can lead to cancer. A defect of the immune system can mean that the body cannot destroy cancer cells that form.

We have come a long way in our understanding of cancer, but the more we learn the more we realize the complexity of the subject. There is hope that we may be able to apply some of our knowledge to further reduce the great toll of human life extracted by this great scourge of humanity.

PROBLEMS AND QUESTIONS

1 Match the following terms with appropriate definitions.

definitions

(1) A malignant neoplasm
(2) The proportion of persons who express a certain gene they carry
(3) A trait that is induced by environmental factors and is identical to a trait caused by genes in other cases
(4) The degree of expression of a specific gene
(5) A measure (often a percentage) of the degree of similarity of paired individuals (e.g., twins) for a specific trait
(6) A chemical or form of radiation capable of inducing cancer
(7) Offspring of the same parents; i.e., brothers and sisters

terms

(a) carcinogenic agent
(b) benign tumor
(c) expressivity
(d) phenocopy
(e) discordance
(f) penetrance
(g) siblings
(h) concordance
(i) cancer
(j) correct response not given

2   Study the following data concerning concordance between twins. What do these data imply about the relative roles of heredity and environment in determining the conditions listed?

	Concordance Values in Percentages	
Condition	Monozygotic Twins	Dizygotic Twins
tuberculosis	53	22
spina bifida	72	33
clubfoot	23	2
measles	95	87

3   Incomplete penetrance creates what difficulties for a geneticist attempting to determine the mechanism of inheritance for a particular trait?

4   A simple expression of the relationship between nature and nurture in determining many traits is summarized in the following statement (choose correct alternatives): (Heredity/Enironment) _____ determines the upper limit of one's development and (heredity/environment) _____ determines how closely this upper limit is approached.

5   When expressivity for a certain trait is so variable that 20 percent of the individuals who have the genotype for the trait do not express it, we can say that the trait has only 80 percent _____.

6   What factors can account for the phenomenon described in Question 5?

7   Numerous chemicals, including many environmental pollutants, and various types of radiation have been shown to be carcinogenic. What common mechanism may underly all of these various "causes" of cancer?

REFERENCES

AMES, B. N. 1979. Identifying environmental chemicals causing mutations and cancer. *Science* 204:587–93.

CAIRNS, J. 1975. The cancer problem. *Scientific American* 232(1):64–78.

CROCE, C. M., and H. KAPROWSKI. 1978. The genetics of human cancer. *Scientific American* 238(2):117–25.

D'EUSTACHIO, P. 1984. Gene mapping and oncogenes. *American Scientist* 72(1):32–40.

GOTTESMAN, I. I. and J. SHIELDS. 1982. *Schizophrenia: The Epigenetic Puzzle*. Cambridge University Press, New York.

HEMMINGS, G., and W. A. HEMMINGS, Editors. 1978. *The Biological Basis of Schizophrenia.* University Park Press. Baltimore, Maryland.

MARX, J. L. 1978. DNA repair: new clues to carcinogenesis. *Science* 200:518–21.

MARX, J. L. 1984. What do oncogenes do? *Science* 223:673–76.

SCHEINFELD, A. 1965. *Your Heredity and Environment.* J. B. Lippincott Co. Philadelphia.

SPARKES, R. S., H. MULLER, I. KLISAK, and J. A. ABRAM. 1979. Retinoblastoma with 13q$^-$ chromosomal deletion associated with maternal paracentric inversion of 13q. *Science* 203:1027–29.

Population Genetics

DID YOU EVER WONDER HOW THE GREAT DIversity in human populations has come about? For instance, why are the giant Watusis of Africa different from the natives of Japan? According to anthropological findings, if we go back some 2 million years we would find that all human beings were members of one population living in east central Africa. Since then they have spread to all parts of the earth and have formed many different populations, each with distinctive inherited characteristics. What forces have brought about these genetic differences, and what changes might be expected in the future? These are the types of questions that population geneticists seek to answer and with which we shall be concerned in this chapter.

Causes of Population Differences

A **population** is a freely interbreeding group living within a restricted area. The genes carried by all the people in a population make up the **gene pool** of that population. Let us examine some of the forces that have caused divergence in the gene pools of different populations.

Natural Selection

During most of the time that human life has existed on this planet, living conditions have been very harsh and survival has been a real problem. With the many famines, plagues, and great fluctuations in climate, there were probably times when *Homo sapiens* was an endangered species. Only a few of the babies born lived long enough to reproduce, but those few had gene combinations that favored life in the environments in which they lived. Since different regions of the earth have different environments, genes that might have been

253

advantageous in one area could have had no advantage, or even could have been harmful, in another area.

The degree of skin pigmentation is an easily recognized characteristic that varies considerably in different populations. As a rule, people who have lived in regions where the sun's rays are intense have more pigment than those who have lived in regions where the sun has less intensity. The heaviest melanin deposits are found among people whose ancestors lived in such regions as central Africa, the interior of Australia, and parts of India. All of these areas have sparse tree growth and therefore little shade. In the equatorial region of South America, the natives are not so heavily pigmented because they live in a forested area where a canopy of tall trees provides a shield from the direct rays of the sun. The people of the Arab nations of North Africa also have dark skins, but not as dark as those of central Africa. As a rule, Italians, Greeks, and Spaniards of southern Europe are somewhat fairer skinned than Arabs, but still are darker than the Germans, Danes, and Swedes who live in northern Europe. In the extreme part of northern Europe—Lapland in northern Scandinavia—the natives have rather dark skin. They live in open country that has a snow cover much of the year. Reflection off the snow makes the sun's rays very intense at times, as anyone who has been sunburned on the ski slopes can testify.

A protection from sunburn would seem to be a logical factor involved in accounting for these differences in skin pigmentation. Severe sunburn can cause blisters that may become infected. In a primitive society this could lead to death in some instances. Fair-skinned persons are also more likely to have skin cancer as a result of excessive exposure to sunlight.

Still another factor could be involved in determining the degree of skin pigmentation. Vitamin D is produced in the skin during exposure to sunlight. A deficiency of this vitamin results in rickets, an affliction that would have greatly handicapped a person in a primitive society where survival depended upon strength and physical agility. Too much vitamin D can also be harmful, causing the bones to develop improperly. Hence, selection would favor people having just the right amount of melanin to permit penetration of the optimum quantity of ultraviolet rays.

Tanning, which is the variation of the amount of melanin as an accommodation to the variations of the intensity of the sunlight at different seasons, is also adaptive. A tan developed in the sun of summer fades in the winter, thereby enabling the skin to use the sun's feeble rays better during this season. Thus, natural selection would tend to establish the right amount of melanin in the skin of members of a population over many generations.

Reproductive Selection

Mere survival is not enough to establish genes in a population. The survivors must leave some of their genes to descendants if they are going to have any influence on the gene pool of the future. Hence, reproductive fitness is just as important as physical fitness. In many forms of life, male strength is favored. The strongest males drive off their competitors and mate with the most females. In primitive human societies such struggles may have occurred. It is also possible that the more powerful males had an advantage in overcoming the resistance of reluctant females. The genes of males who had many qualities for survival but had low sexual drive or were impotent probably would have perished with their bodies. Likewise, women with disorders of the reproductive system that prevented conception or caused abortions would not have passed along any genes even though they may have been well endowed physically in other respects.

As societies became more complex, shrewdness may have become more important than physical strength. A man who could accumulate wealth in one form or another could buy women for a harem and pass on a disproportionate share of his genes to future generations. Therefore, customs could have resulted in some differences in populations.

Reproductive selection may sometimes result in the establishment of traits with no particular survival value. Women and men with facial features and body forms that are regarded as sexually appealing in a particular society would have an advantage over women and men further removed from the ideals of the society. That ideals of feminine and masculine attractiveness vary considerably in different parts of the world could have been a factor in establishing certain features that characterize population groups today. For example, the amount of body hair on men varies greatly even in areas where climatic conditions are similar. Most Japanese men have sparse body hair, but the Ainu men on islands just north of Japan are among the most hairy men in the world. It is possible that ideals of masculine attractiveness have differed in these two populations, which have remained isolated from one another for a long time.

Cultural Selection

About 10,000 years ago, only a tiny fraction of the time human life has existed on the earth, people discovered ways to cultivate plants and domesticate animals. This enabled them to give up their nomadic life and settle in permanent villages. They then began to find ways to modify their environment to make living easier. This has led to housing with central heating and air conditioning, which can make life pleasant in such extremes as central Alaska and the deserts of the American

Southwest. Also, public health and medical measures have greatly relaxed the fierce pressure of diseases that once played a large part in natural selection. In addition, in many societies the favored person no longer is the one with the greatest brute strength, but rather is the one with the ability to cope with complex cultural conditions. Still, cultures vary considerably in different parts of the world, and these differences continue to be a factor in establishing and maintaining population differences.

Genetic Drift

The desire to move, to find new and better living conditions, seems to be a human characteristic. In all of human history we find accounts of small groups moving away from established populations to distant lands. For instance, many of the islands of the South Pacific are populated by descendants of Tahitians who traveled over thousands of miles of open water in double canoes to found new colonies. The proportions of genes in any such small migrating group are not necessarily representative of the population from which they came. The small migrating group may thus reflect a **sampling error** with respect to the larger "parent" population. Hence, as the "daughter" population increases in numbers in its new home, its gene pool may become quite different from that of the parent population. Differences in gene frequencies between parent and daughter populations may be enhanced because of different evolutionary pressures operating on the gene pools of the two populations in their distinct and separate environments. What has been described in this paragraph is known as the **founder principle.** Directly related to the founder principle is the concept of **genetic drift,** the random fluctuation of gene frequencies due to sampling errors. Genetic drift can occur in any population, but it is especially evident in very small populations, because the probability of sampling error is greater with such small groups.

The Maoris, natives of New Zealand, are descendants of Tahitians who came to that part of the world about 1400 years ago. They have a much lower frequency of the gene for the A blood antigen than the Tahitians of today. The thirty or so people who crowded into those canoes possibly included a smaller percentage with type A than was the average for the Tahitian population of that time. Also, during the first generations, those with type A blood might not have averaged as many children as those with other types. Additional factors, such as immigration from other areas at later times, could also have had an influence.

Genetic drift can also occur in established populations as a result of catastrophe. About 6 percent of the people living on the island of Pingelap, one of the eastern Caroline Islands in the South Pacific, have complete congenital **achromatopsia,** a recessive trait characterized by abnormal cones of the retina. Cones are used primarily in bright light and are color sensitive, so people with this defect cannot stand exposure to bright light and are totally colorblind. The mutant gene involved is relatively rare in the world as a whole, so why should it have such a high frequency on this island? Records show that a great hurricane swept over this island around 1780, leaving about nine surviving males. Evidently, a high percentage of the survivors had this gene.

Even when a small group migrates into an area populated by others, genetic drift can be of significance, provided the migrants do not intermarry with their neighbors. Bentley Glass (see References) studied an interesting population of Dunkers, a religious sect living in Franklin County, Pennsylvania. Dunkers are descendants of the Baptist Brethren who lived in the Rhineland region of Germany near Krefeld. In 1719 a group of twenty-eight migrated to America. They had strict rules against marriage outside their group, and so have remained isolated genetically. There has been some migration outward to, but practically no migration into the group from, the surrounding area. Analysis of their blood types is presented in Table 15–1. It appears that genetic drift is the reason for the higher percentage of type A individuals in the Dunkers than is found in the German people from whom they descended. There seems to have been no inflow of genes from surrounding Pennsylvanians as shown by the fact that the Pennsylvanians have even fewer type A persons than the Germans.

TABLE 15–1. Genetic drift effect on blood types of Dunkers

Population	Percentage of ABO Groups			
	O	A	B	AB
Dunkers	35.5	59.3	3.1	2.2
Rhineland Germans	40.7	44.6	10.1	4.7
Pennsylvanians	45.2	39.5	11.2	4.2

Differential Mutation

Some genes mutate very rarely and may become established in a single locality. The genes for the blood antigens are possible examples. The original human beings may all have been type O, but perhaps in western Europe a mutation to A occurred. This mutation, although it has spread over most of the earth in the thousands of succeeding generations, is still most frequent around its point of origin. The mutation to produce the B antigen seems to have occurred much later in southern China. This gene has also been spread, but not as far because of its more recent origin. Both the American Indians and the Australian aborigines, who seem to have Asian origins, are practically free of the gene for B. These populations must have migrated before the mutation became widespread in Asian populations.

Alternate Pathways of Selection

There is not necessarily one gene combination that is best for a particular environment. Several ways exist in which adaptation can be achieved, and chance may play a role in determining just which pathway will be followed. Resistance to malaria, for instance, has been accomplished in central Africa by selection favoring the gene for sickle-cell anemia, but in other areas a different gene accomplishes the same thing. In Italy, Greece, and other countries bordering the Mediterranean Sea, as well as in parts of Asia, the gene for Cooley's anemia, or thalassemia, confers the resistance. Individuals homozygous for this gene have severe anemia and only through continued transfusions can they be saved from early death, but the heterozygotes are highly resistant to malaria. (In the United States about one person in twenty-five with Italian or Greek ancestry carries this gene. Chance may have been responsible for the establishment of this gene rather than the gene for sickle-cell anemia.) In Sicily and Sardinia the gene for G6PD deficiency (see Chapter 6) has a high frequency and protects against malaria. No doubt, many population differences can be accounted for by such chance selection along alternate pathways.

Gene Flow Through Migration

The gene pool of most populations is constantly being altered because of gene flow from other populations. The Italian people of today are not counterparts of the ancient Romans. Neither are modern Greeks or Egyptians like their ancient ancestors.

Gene flow can be easily noted in contact zones. The French living on the borders of Spain show many Spanish characteristics. In addition, migrants to new lands may remain in isolated groups at first, but in time the barriers break down and their genes blend with those of the people in their new country. The Swedes who migrated to Minnesota at first kept their language, customs, and religion and married primarily within their own group. Associations in public schools and in social activities, however, have resulted in intermarriage with people from other cultures. In time there will be no distinction between this group and others in that region of the country. In Israel, where Jewish people from different regions have come together, it is evident that German Jews differ from the Russian Jews, and both groups differ from the Spanish Jews. In spite of religious restrictions, there has been gene flow from other kinds of people into the Jewish population.

The more a migrating group differs from those around it, the slower the mingling of genes will be. In the United States, Japanese immigrants blend more slowly than do German immigrants because they are more different from the majority of Americans. Blacks, brought to this country some three hundred years ago, have blended slowly because of their great difference in skin color. Yet, studies by Bentley Glass and C. C. Li (see References) indicate that blacks in America, on the average, have about 30.46 percent of their genes from white ancestors. Other, more recent studies, suggest that this percentage is too high and may be of the order of 22–28 percent. This amount of flow has occurred in about ten generations and will no doubt accelerate as the differences between the two groups become less distinct.

Differential reproduction of immigrated groups can play a significant role in changing a gene pool. Jamaica, for instance, was once inhabited only by Indians, but when the British came they brought many black Africans as slaves to work on sugarcane farms. These Africans had a high rate of reproduction, so the population of Jamaica now is primarily black. In addition, wars and disease played an important part in reducing the Indian population. A similar situation exists on Fiji. There the native people are black and have had a rather low rate of reproduction, which has kept their population stable. At the beginning of this century the British imported Asiatic Indians to work in the sugarcane fields. These immigrants had a higher reproductive rate than the native Fijians, and they and their descendants now are about equal in number to the natives. The groups are beginning to intermarry more and more, however, and in time the population will probably become somewhat homogeneous, with the gene pool quite different from what it was before this century.

Invasion
Invaders, both peaceful and forceful, invariably scatter some of their genes along their routes. Each incursion into a country by people of a different genetic background is bound to alter the gene pool. The many war brides brought back from invaded countries, such as Germany, Japan, Korea, and Vietnam, will all play a part in alteration of the American gene pool.

The influence of invasion is so powerful that it is possible to trace the route of invading armies by the gene pool of the populations along the way even many generations later. The route taken by the invading Mongols into Europe hundreds of years ago can still be plotted by the percentages of the B antigens in the European populations today. Certain isolated regions escaped the invaders. The Basques, on the northern coast of Spain, protected from the invaders by high mountains, probably possess a gene pool like that of the ancient Europeans. They have only 1.1 percent B antigens, while other regions of Spain have about 9.2 percent. A similar area is found in the Caucasus area of Georgia and Armenia, which are parts of the Soviet Union. The distri-

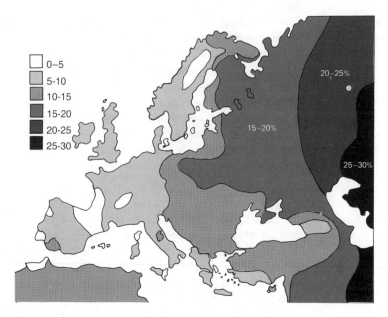

FIGURE 15–1. Distribution of B type blood in Europe and part of Asia, showing how the allele for the B antigen has spread from its origin in southern China.

bution of the B antigen in Europe and part of Asia is shown in Figure 15–1.

Another good example of the influence of invasion shows up in a survey of the distribution of blood antigens in Great Britain. A cline, or gradual change, from high A in the south to high O in Scotland can be noted. The original Britains were probably mostly type O, but invaders from across the channel brought in many genes for the A antigen. In Ireland also, the cline is from high A in the east to high O in the west. Invaders from England, after they had acquired considerable A, brought their genes with them as they tried to subdue the rebellious people across the Irish Sea.

Heterozygote Superiority

Harmful genes are not evenly distributed in the populations of the world. Tay-Sachs disease is common among the Ashkenazic Jews, but very rare among others. Sickle-cell anemia is common among blacks, but very rare among others. Cystic fibrosis is common among whites, but much less common among others. Table 15–2 shows the uneven distribution of a number of such harmful traits.

Heterozygote superiority could account for some of this fluctuation. Individuals who are heterozygous for the gene for sickle-cell anemia

TABLE 15–2. Fluctuations in frequency of harmful traits

Trait	Population	Frequency per 100,000
spina bifida	Irish	426
	Japanese	21
pyloric stenosis (constriction of the pyloric valve from the stomach)	U.S. Caucasians	159
	Hawaiians	5
Tay-Sachs disease	U.S. Jews	62
	other U.S.	near 0
cystic fibrosis	U.S. Caucasians	26
	Japanese	1
sickle-cell anemia	U.S. blacks	200
	U.S. Caucasians	near 0

Data from N. E. Morton.

resist malaria much better than those who are homozygous normal. Thus, selection favors the carriers of the gene, and it can reach a high frequency in a region where malaria is prevalent. In other areas, selection is all against the gene by elimination of the homozygotes. Perhaps many of the other fluctuations are due to a similar advantage of the heterozygote in a particular environment. Some have suggested that the Ashkenazic Jews have a high incidence of Tay-Sachs disease because those heterozygous for this gene may for some unknown reason have been better able to survive the harsh ghetto existence forced upon them in eastern Europe. Some evidence suggests that T/t heterozygotes are slightly more fertile than T/T homozygotes; this could account for maintaining the t allele in the Jewish population at a relatively high frequency.

Polymorphism of Gene Pools

In a particular environment you might expect one allele at a locus on a chromosome to be more advantageous than other alleles at this locus. If selection is the primary force in establishing a gene pool, this allele should be established at a very high frequency while the other alleles would be eliminated or maintained at a very low frequency consistent with continued mutation. However, two or more alleles are sometimes maintained at a relative high frequency. This is known as **genetic polymorphism.** The many varieties of eye color and hair color among Caucasian Americans are a result of polymorphism of genes involved in producing these traits. Let us examine some of the factors that may be involved in establishment of genetic polymorphism.

Transient Polymorphisms

Polymorphism of some genes can be established when the environment undergoes a marked change or when a population migrates to a new region where the environment is different. A gene that may have been relatively rare because it was disadvantageous may now assume an advantage and, through selection, its frequency will increase. For quite a few generations this gene, as well as its formerly favored allele, will exist at high frequencies, but this is a transient condition. In time the allele that is now less advantageous will decrease until it is at a very low frequency. When an allele frequency drops below about one percent, geneticists no longer consider a polymorphism to exist. Genes that had an advantage in a stone-age culture have, no doubt, declined in favor of alleles that are more compatible with cultural conditions in modern times. The blacks in America still have genetic polymorphism of the alleles for hemoglobin S and hemoglobin A, but the allele for S hemoglobin should be gradually declining since malaria has been conquered in the United States.

Balanced Polymorphisms

Sometimes two or more alleles remain at relatively high levels in a population over a long period of time because there is a balance between the advantageous and disadvantageous actions of the alleles. The gene for hemoglobin S became established in Africa at a high frequency because of the opposing forces of selection. It will remain balanced at that level as long as malaria exists and there is a selective advantage to the heterozygote.

Sometimes medical techniques maintain a balanced polymorphism of some harmful genes. Genes conducive to diabetes are relatively frequent in the people of the United States even though they can be very harmful. This is especially true of the Pima Indians of Arizona. About 50 percent of these people develop diabetes between the ages of twenty and fifty years. They have a background of life in a very harsh desert environment where the food supply was very irregular; the genes conducive to diabetes might have permitted them to make maximum use of the scant nourishment available to them. In the present culture, however, most of them are regularly eating carbohydrate-rich diets, and so many develop diabetes. Genes that favored full use of the few carbohydrates available now leave them with an excess of carbohydrates in the blood. This would probably be a case of transient polymorphism except that through the use of insulin and other medical techniques, the diabetics can lead comparatively normal lives and contribute their share of genes to the gene pool. Hence, the high incidence of diabetes in this group will probably continue.

In primitive societies, seasonal selection can also maintain two alleles in a balanced polymorphism. In areas of extreme climatic fluctuation, one allele may have an advantage in cold weather and another in warm weather. Hence, selection favors one in winter and the other in summer, and both are maintained at relatively high levels.

Sexual dimorphism has been established in many forms of life, including humans. Genes that have a value in females may have no value, or actually may be disadvantageous, in males. Sex limitation of some of these genes has been established. A rather wide-spaced pelvis is an advantage to a woman carrying a fetus and giving birth, but the wider spacing of the leg bones that results would slow down a man as a hunter. Hence, the expression of the genes for this trait has become sex limited to females. Sexual selection may also play a part in establishing sexual dimorphism, as was discussed earlier in this chapter.

Polymorphism of the Blood Groups

Most human populations have polymorphism for the three alleles for the ABO blood groups. How are these alleles maintained in the population? If one has any physiological advantage over the other two, then it seems as if it would become predominant through selection. Some studies indicate that duodenal ulcer is about 40 percent more prevalent in people with type O blood than in those with other types. On the other hand, cancer of the stomach appears to be more common among people with type A blood. A study in China showed that people with type O are more resistant to pulmonary tuberculosis. A balance between the various advantages and disadvantages of the blood types might be a factor in maintaining polymorphism.

Maternal-fetal incompatibility could also be a factor in light of the recent findings that such incompatibility can be a major cause of fetal death. Fetuses that have type O blood would be favored overall because they would not react to either anti-A nor anti-B antibodies from the mother, yet this advantage might be mitigated by other forces that would favor the A or B antigens. Heterozygote superiority (heterosis) could also be involved. Some studies indicate that fetuses that are heterozygous for any of the blood group genes have some advantage over homozygotes. Detailed experiments in rats, mice, and cattle confirm this possibility.

Antigenic similarity of blood antigens to the antigens of parasites, such as worms and bacteria, could be another force involved in maintaining polymorphism. Some of these parasites may have antigens similar to A and B, and thus would not be rejected as readily in a host with these antigens. For instance, if a hookworm had an antigen similar to A and it infected a person who had type A blood, no antibodies against it would be produced because the B lymphocytes that could produce them were eliminated during fetal life. Those people who had type O or B blood, however, could produce the antibodies; in fact, they would already have anti-A and might reject the infection. On the other hand, the bacterium that causes cholera might have antigens similar to B antigen, so those persons with types A and O blood would be favored in regions where cholera is prevalent.

Physiological functions could also be involved. Studies in sheep and other animals show that certain blood antigens in red blood cells made these cells more efficient in transporting vital enzymes and minerals to the interior of the cells.

A combination of these mechanisms and other forces still unknown could maintain the alleles for different blood groups in a polymorphic state.

Gene Frequencies in Populations

The frequency of specific genes in a population can be determined by mathematical calculations based on the numbers of individuals who express these genes. The **Hardy-Weinberg principle** was formulated in 1908 by an English mathematician, G. H. Hardy, and a German physician, Wilhelm Weinberg. It holds that in a stable and randomly breeding population, allele frequencies would be in equilibrium, and that the frequency of a recessive gene could be determined from the percentage of individuals who express the gene in a representative sample. For example, a gene that would seem to have no selective advantage or disadvantage is the one for the ability to taste the chemical phenothiocarbimide (PTC). Some people get a very bitter taste from a piece of paper impregnated with this chemical, but others get no taste at all. Suppose you test hundreds of people in a population and find that 64 percent are tasters and that the remaining 36 percent are nontasters. The nontasters are homozygous for the recessive allele t/t. From this information we can determine the percentage of tasters who are homozygous T/T and heterozygous T/t.

Allow q to represent the frequency of the gene t, and p to represent the frequency of the dominant allele T. The frequency of nontasters would be q^2, and we can get the value of q by taking the square root of q^2. The frequency of p would be $1 - q$. Determination of the values would be as follows (see also Figure 15–2 for a graphic representation of this information):

$$q^2 = 0.36 = \text{observed frequency of nontasters } t/t$$

$$q = \sqrt{.36} = 0.60 = \text{frequency of } t$$

$$p = 1.00 - 0.60 = 0.40 = \text{frequency of } T$$

$$p^2 = (0.40)^2 = 0.16 = \text{frequency of homozygous tasters } T/T$$

$$2pq = 2 \times 0.60 \times 0.40, \text{ or } 1.00 - (0.16 + 0.36) = 0.48$$

$$= \text{frequency of heterozygous tasters } T/t$$

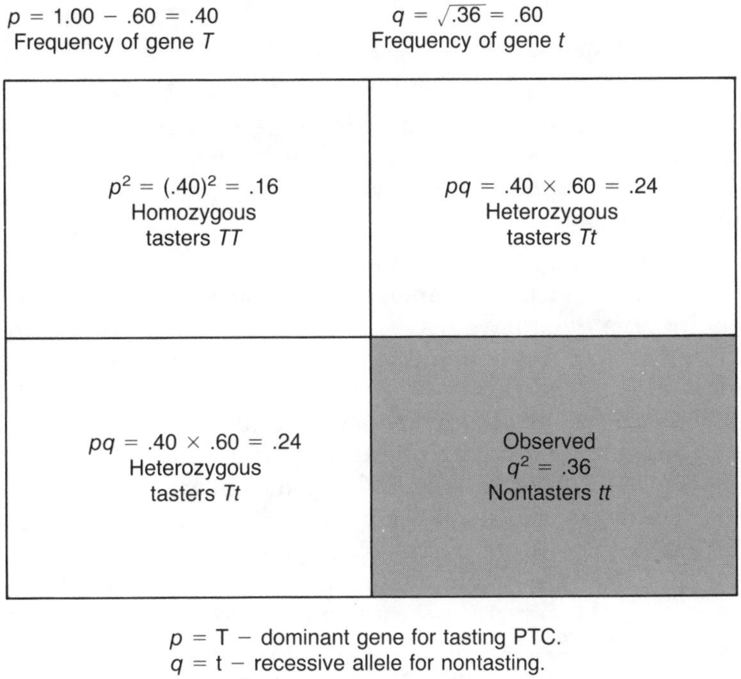

$p = 1.00 - .60 = .40$
Frequency of gene T

$q = \sqrt{.36} = .60$
Frequency of gene t

$p^2 = (.40)^2 = .16$ Homozygous tasters TT	$pq = .40 \times .60 = .24$ Heterozygous tasters Tt
$pq = .40 \times .60 = .24$ Heterozygous tasters Tt	Observed $q^2 = .36$ Nontasters tt

$p = T -$ dominant gene for tasting PTC.
$q = t -$ recessive allele for nontasting.

FIGURE 15–2. Graphic method illustrating the use of the Hardy-Weinberg principle to determine gene frequencies. Since 36 percent of a population sample cannot taste PTC, and since nontasters are homozygous t/t, the values of q, p, p^2, and $2pq$ can readily be obtained.

This principle can be used to determine whether specific genes have a selective advantage or disadvantage. For instance, we assume that the M and N blood antigens are neutral, but there could be some effect that we do not know about. We can find out through the Hardy-Weinberg principle.

About 20 percent of the people of Sicily are type N. These people are homozygous L^N/L^N. From this we can calculate the number of people who would be expected to be type M and MN:

$$q^2 = 0.20 = \text{observed frequency of type N, genotype } L^N/L^N$$
$$q = \sqrt{.20} = 0.45 = \text{frequency of gene } L^N$$
$$p = 1 - 0.45 = 0.55 = \text{frequency of gene } L^M$$
$$p^2 = (0.55)^2 = 0.30 = \text{frequency of type M}$$
$$2pq = 2 \times (0.55 \times 0.45) = 0.50 = \text{frequency of type MN}$$

In this case we can easily identify the heterozygote, so we can compare the actual distribution with the predicted distribution:

	M	**MN**	**N**
Calculated percentage	30	50	20
Observed percentage	32	48	20

The calculated and observed percentages of heterozygotes are so similar that we can feel reasonably sure that mating is random with regard to this trait, and there is no survival advantage for any of the three possible genotypes.

As an example of genes that have a selective advantage, let us consider the gene for **sickle-cell anemia.** Since most babies who are homozygous for this gene have severe anemia and often die within a few years after birth, it might appear that selection would greatly favor the dominant allele. The Hardy-Weinberg principle will tell us if this is true. Sickle-cell anemia has a frequency of about 4 percent of the newborns in Nigeria. The square root of this figure indicates a frequency of 20 percent for the Hb^S allele. Hb^A would thus have a frequency of 80 percent. The expected frequency of homozygous normal individuals would thus be $p^2 = (0.8)^2 = 0.64 = 64$ percent, and the expected frequency of heterozygous individuals would be $2pq = 2(0.8)(0.2) = 0.32 = 32$ percent. We can make an actual determination of these carriers by exposing the blood to reduced oxygen concentration because this causes sickling of some cells. Or we can use electrophoresis to separate out the two kinds of hemoglobins. Such tests show that actually 46 percent of the population is heterozygous. This is significantly higher than the expected 32 percent, and reflects the selective advantage of the heterozygote in resisting malaria. No doubt many of the homozygous normals died of malaria and those homozygous for the abnormal allele died of sickle-cell anemia.

Genes of Lesser Frequency

When a recessive trait is relatively rare, percentages involve the use of many decimals and it is sometimes more convenient to use fractions. For instance, about 1 in 10,000 Caucasians born in the United States has phenylketonuria or PKU (see Chapter 9). This is only 0.01 percent of the population, or 0.0001. It is simpler to use 1/10,000 as the gene frequency in calculations. The various frequencies would be as follows:

$$q^2 = 1/10,000 \text{ so } q = 1/100 = \text{frequency of gene } p$$
$$1 - 1/100 = 99/100 = \text{frequency of allele } P$$
$$(99/100)^2 = 9801/10,000 = \text{frequency of } P/P$$
$$1 - 9802/10,000 = 198/10,000 = \text{frequency of } P/p$$

The Significance of the Hardy-Weinberg Principle

The Hardy-Weinberg principle suggests that a population is in an **equilibrium** with respect to a certain gene locus. That is, for generation after generation, the values of $p =$ frequency of A and $q =$ frequency of a do not change, nor do the frequencies of the genotypes A/A, A/a, and a/a. To be in such a state of equilibrium with respect to a gene, a population must mate at random, with neither mutation nor selection favoring one allele over the other. That is, all genotypes—A/A, A/a, and a/a—must have equal survival and reproductive fitness. If mutation occurs, the frequency of $A \longrightarrow a$ mutation must equal the frequency of $a \longrightarrow A$. Furthermore, to remain in a Hardy-Weinberg equilibrium the population cannot experience any major immigration nor emigration that will alter the frequencies of A and a.

Data in Tables 15–3 and 15–4 show that for a hypothetical, random mating population in which p = frequency of A = 9/10 and q = frequency of a = 1/10 and p^2 = 81/100 A/A, $2pq$ = 18/100 A/a, and q^2 = 1/100 a/a, such an equilibrium exists. Shown in Table 15–3 are the various possible matings in such a population and the frequencies of these matings if random mating occurs. Shown in Table 15–4 are the frequencies of A/A, A/a, and a/a offspring produced, assuming that all matings are equally fertile (i.e., in this example each mating produces one offspring). Note that the bottom line in Table 15–4 shows that after one generation of random mating, the population is still in a Hardy-Weinberg equilibrium with 81 percent A/A, 18 percent A/a, and 1 percent a/a.

Gene Equilibrium of Harmful Genes

Outgo by Genetic Death

It might seem as if selection would gradually eliminate harmful genes from a population. Any gene that is harmful will tend to reduce viability and/or ability to reproduce. Thus, many harmful genes are eliminated from a population each generation. This may even be a 100 percent elimination if none of the people who express the trait live long

TABLE 15–3. Frequencies of different kinds of matings in a random mating population with p = 0.9 and q = 0.1

men \ women	81/100 A/A	18/100 A/a	1/100 a/a
$\dfrac{81}{100}$ A/A	$\dfrac{6561}{10,000}$ $A/A \times A/A$	$\dfrac{1458}{10,000}$ $A/A \times A/a$	$\dfrac{81}{10,000}$ $A/A \times a/a$
$\dfrac{18}{100}$ A/a	$\dfrac{1458}{10,000}$ $A/a \times A/A$	$\dfrac{324}{10,000}$ $A/a \times A/a$	$\dfrac{18}{10,000}$ $A/a \times a/a$
$\dfrac{1}{100}$ a/a	$\dfrac{81}{10,000}$ $a/a \times A/A$	$\dfrac{18}{10,000}$ $a/a \times A/a$	$\dfrac{1}{10,000}$ $a/a \times a/a$

TABLE 15—4. The consequences of one generation of random mating in a population in a Hardy-Weinberg equilibrium with $p = 0.9$ and $q = 0.1$

Frequency of Matings*	Matings	Results of Matings		
		A/A	A/a	a/a
$\frac{6561}{10,000}$	A/A × A/A	$\frac{6561}{10,000}$		
$\frac{2916}{10,000}$	A/A × A/a	$\frac{1458}{10,000}$	$\frac{1458}{10,000}$	
$\frac{162}{10,000}$	A/A × a/a		$\frac{162}{10,000}$	
$\frac{324}{10,000}$	A/a × A/a	$\frac{81}{10,000}$	$\frac{162}{10,000}$	$\frac{81}{10,000}$
$\frac{36}{10,000}$	A/a × a/a		$\frac{18}{10,000}$	$\frac{18}{10,000}$
$\frac{1}{10,000}$	a/a × a/a			$\frac{1}{10,000}$
Totals		$\frac{8100}{10,000}$	$\frac{1800}{10,000}$	$\frac{100}{10,000}$
Frequencies		$\frac{81}{100}$	$\frac{18}{100}$	$\frac{1}{100}$

*Frequencies in this column were summarized from Table 15–3.

enough to reproduce, or if they are sterile. Selection eliminates dominant harmful genes rather quickly. If 50 percent of those who express the trait have a genetic death, then half of the genes are eliminated each generation. If all have a genetic death, the gene is never propagated.

Selection cannot operate against recessive genes until they become homozygous. In a stable population, about the same number of harmful recessive traits appear each generation. Among Caucasians, about 1 in 20 carries the recessive gene for cystic fibrosis in the heterozygous state. About 1 marriage in each 400 is between two such carriers ($1/20 \times 1/20 = 1/400$). Since about one-fourth of the children of two carriers become homozygous recessive, about 1 baby out of every 1600 born has cystic fibrosis. Until recent times all such children perished long before maturity, so selection was 100 percent against the trait. The frequency does not decline, however, because new genes for cystic fibrosis are being added each generation through mutation.

Input by Mutation
Genes mutate at rare intervals and since most mutations are harmful, this input tends to counterbalance the elimination by genetic death. In a stable population an equilibrium is established. This equilibrium can

be upset, however, when there is a change in the rate of input or a change in the outgo. Unfortunately, for the first time in history, things are happening today that both increase the input and decrease the outgo of harmful mutations.

Genetic Pollution

Recently we have become greatly concerned about the pollution of our environment, and much money is being spent to try to prevent the harmful effects of such pollution on our health and well-being. Comparatively little concern is being expressed, however, for another kind of pollution that will affect not only those of us living today, but all future generations who are dependent on the pool of genes carried by those living today. This pollution is genetic pollution.

Modern sanitation, medical treatment, surgery, and preventive medicine are saving many individuals whose genes formerly would have caused genetic death. Each time a child is saved by open-heart surgery, we perpetuate any genes that may have been involved in causing the abnormality. Each time a woman and her baby are saved by a Caesarian operation, we make possible the continuation of any genes that may have made a normal birth impossible. Each time a child is saved from PKU by diet, two genes are preserved that otherwise would have been eliminated. Such reduction of the natural methods of purifying the gene pool will allow more harmful genes to remain in the pool. The increase will be slow because most such genes are carried by heterozygous persons, but over a long time an increase in harmful traits can be significant.

In addition, we are being exposed to mutagenic agents today that can increase the input of harmful mutations. X rays and radioactive isotopes are used extensively for the diagnosis and treatment of various afflictions, but such radiation is mutagenic. In our polluted environment we are exposed to mutagenic agents in air, focd, and water. At the present rate, this addition of harmful mutations is small compared to the natural addition, but if continued for many generations, it could become significant.

Possible Solutions

What can be done to prevent a deterioration of our gene pool? Will it continue until all the people in some distant future generations will suffer from one or more serious genetic defects and can be kept alive only through special treatment? Some have suggested that human selection be substituted for natural selection to prevent such deterioration. When we save persons from a serious genetic disease, does that necessarily give them a right to perpetuate the genes for the defect? Some would say that we can at least advise such individuals against having children, while others would say that not having children should be mandatory. For very harmful traits, we could even extend this human selection to the carriers. We now require tests for venereal diseases before issuing a marriage license, and this could be extended to genetic diseases as well. Compulsory restriction, however, is against our principles of individual freedom and there is some effort to use voluntary means. Jewish people, for instance, are instigating a program of testing for carriers of the gene for Tay-Sachs disease. In some cities, such as Baltimore, they are asking all who prove to be carriers to refrain from having children. If all cooperated, this horrible disease, which afflicts about 1 Jewish baby out of each 3600, could be wiped out in one generation. A similar program is being tried for sickle-cell anemia in blacks. Since about 1 in 10 American blacks is a carrier of the Hb^S gene, this would mean quite a sacrifice for one generation, but then no longer would so many have to suffer from this affliction.

In addition, we can take steps to prevent introduction of excessive numbers of harmful mutations into the gene pool. Exposure to radiation from medical treatment and diagnosis should be kept at a minimum consistent with the health of the present generation. Release of radioactive wastes from nuclear generators and other industrial plants should be kept as close to zero as practical. The use of possible mutagenic agents such as pesticides, herbicides, and food additives should be restricted. As carriers of the germ plasm for all future humankind, we have a certain responsibility to see that it is not unduly polluted by what we do today.

Efficiency of Selection

Selection against recessive traits reduces the gene frequency rather rapidly when the trait is common, but the efficiency of selection declines as it becomes less common, because with fewer genes in the pool, more of them are carried by heterozygotes and these are not selected against.

Complete Selection

To illustrate the efficiency of complete selection against a recessive trait, let us consider an isolated population in which 16 percent have blue eyes. Considering blue eyes as resulting from a simple recessive allele of the gene for brown eyes, this would mean that about 40 percent of the genes at this locus are for blue eyes (the square root of 0.16). Suppose some paranoid dictator gets control of the country and decides that blue-eyed persons are a menace to society and should be eliminated. Instead of using the firing squad, the dictator decides to sterilize all people with blue eyes. How fast would this complete selection against a recessive trait reduce the incidence of that trait in the population? A simple equation gives the answer:

$$q_n = \frac{q_o}{1 + nq_o}$$

where n = number of generations, q_o = original gene frequency (0.40 in this case), and q_n = gene frequency after n generations.

If we apply this to one generation we get

$$q_1 = \frac{0.40}{1 + (1 \times 0.40)} = 0.286 \text{ or } 28.6\%$$

Blue-eyed children born: $0.286 \times 0.286 = 0.082$ or about 8.2%

The reduction from 16 percent to 8.2 percent for the first generation might fool us into thinking that blue eyes could be eliminated very rapidly. If the policy was continued, however, blue-eyed babies would still be born even after many generations. After four generations the numbers would be

$$q_4 = \frac{0.40}{1 + (4 \times 0.40)} = 0.154 \text{ or } 15.4\%$$

Blue-eyed children born: $0.154 \times 0.154 = 0.0237$ or 2.37%

After ten generations it would be

$$q_{10} = \frac{0.40}{1 + (10 \times 0.40)} = 0.08 \text{ or } 8\%$$

Blue-eyed children born: $0.08 \times 0.08 = 0.0064$ or 0.64%

Figure 15–3 shows how the efficiency of selection decreases for each generation as the number of genes becomes fewer. More and more of the genes are carried by the heterozygotes and the elimination of homozygotes only reduces the frequency very slowly as the number of "undersirable" genes becomes small.

Incomplete Selection

In natural selection many harmful genes are not completely eliminated when homozygous, but there is some elimination, which reduces the frequency of the gene, but to a slower degree. This incomplete selection can be included in the equation by inserting the letter k to indicate the percentage that is eliminated each generation:

$$q_n = \frac{q_o}{1 + knq_o}$$

If natural selection eliminates 50 percent of those who express a recessive phenotype, the results after one generation will be

$$q_1 = \frac{0.40}{1 + (0.5 \times 1 \times 0.40)} = 0.333 \text{ or } 33.3\%$$

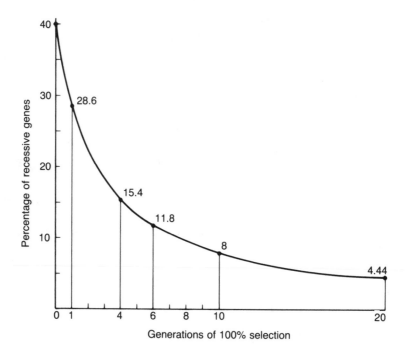

FIGURE 15–3. Effectiveness of 100 percent selection against those who express a recessive trait. As the genes become less common, the efficiency of selection decreases.

The frequency of the gene has been reduced to only 33.3 percent as compared to 28.6 percent when there is 100 percent selection. After four generations of selection at this 50 percent rate, the genes for blue eyes would be 22.2 percent compared to 15.4 percent, and after ten generations it would be 13.3 percent compared to 8 percent.

Consanguineous Matings

Close inbreeding tends to bring out recessive traits because the parents can have the same recessive genes in common. For instance, your chance of carrying the gene for enzyme-deficient albinism is about 1 in 71. If you marry a nonrelative, the chance that he or she will also carry this gene is also 1 in 71, so your chance of having an albino child is $1/71 \times 1/71 \times 1/4 = 1/20{,}164$. If you marry a first cousin, however, the odds are increased. Your chance is still 1/71, but if you carry it, the chance that your cousin will carry it is 1/8. This is the chance that a common grandparent will pass any one gene to both of you. The odds for albinism in a child rises to $1/71 \times 1/8 \times 1/4 = 1/2272$, a 9-fold increase. For genes that are more common, the increase is less. Your chance of carrying the gene for cystic fibrosis is 1 in 20, so your chance of having a child with this condition is 1/1600 from a nonrelative, but is 1/640 from a first cousin. This is only a 2.5-fold increase.

Most harmful recessives are relatively rare, but there are many of them. Hence, the number of babies with genetic defects increases greatly with the degree of consanguinity of the parents. This can be detected in a small population living in an isolated region where there is a restricted choice of mates. Examples would include certain isolated rural populations in northern Japan and the Amish of eastern Indiana.

On occasion, even closer genetic relationships give rise to babies. Incestuous relationships, such as between a father and his daughter or between a sister and her brother, may result in children. The odds for defects in such babies jump greatly. For albinism it would be 1/568, a 4-fold increase over the chance from first-cousin marriage and a 35-fold increase over that for children of nonrelated persons. When we consider that each person carries about two recessive lethals, on the average, the chance that a child of an incestuous mating will die from such a lethal is 1 in 4. A daughter has one-half of the genes of the father, so for one gene carried by the father the chance would be $1/2 \times 1/4 = 1/8$. Multiply this by two, for the two lethal genes, and we find that the chance of a child receiving either lethal in the homozygous state is 1/4. It would be the same for brother-sister matings. In addition, there are many harmful recessives that are not lethal, and some of these would be expressed in the children who lived.

One final word should be mentioned about consanguineous matings. While they are likely to be bad for the children of such matings, they are actually good for the gene pool, since they tend to bring out harmful recessives and selection can purge them from the pool. If continued over a number of generations, the number of defects may actually become less. In the Ptolemaic dynasty of ancient Egypt, brother-sister marriage was the custom for many generations to prevent mingling of "royal blood" with that of commoners. At first there must have been some horrible results, but in time the group was purified to such an extent that no great number of defects appeared. Cleopatra was the product of such a marriage and, from all accounts, she seems to have had no serious genetic defect. She was married to her brother, but preferred the amorous advances of visiting Romans and never lived with him as a wife.

Thus, we see that population genetics has many ramifications and gives us a better understanding of our history and the development of populations on the earth today.

PROBLEMS AND QUESTIONS

1   There is some evidence that the incidence of Huntington's Disease (HD) is somewhat higher in Indiana than elsewhere in the United States. This may be attributed to the fact that the early settlers in the state included a disproportionate number of HD victims whose

descendents are still found in Indiana. The situation just described appears to be an example of what genetic phenomenon?

2   *Genetic load* is a term sometimes applied to the sum total of deleterious or harmful genes occurring in the gene pool of a population. What factors in contemporary society seem to be working to increase the genetic load of humankind?

3   Any practice that increases the genetic load may be called *dysgenic*. What developments in modern medicine can be expected to have a dysgenic effect on the human gene pool?

4   What conditions must be met if a population is to be in a Hardy-Weinberg equilibrium with respect to the alleles of a given gene locus?

5   If 16 percent of the U.S. white population is Rh-negative (d/d), what percent should be homozygous (D/D) for the Rh-positive allele and what percent should be heterozygous (D/d)?

6   Would you regard a population consisting of 33 percent A/A, 58 percent A/a, and 9 percent a/a to be in a Hardy-Weinberg equilibrium? Justify your answer.

7   According to an article in *Science* in 1978, the frequency of a genetic defect called ataxia telangiectasia (AT) is about 1 in 40,000. Individuals with AT are homozygous recessive (t/t) for an autosomal allele. Use this information to calculate the frequency of t and the frequency of T. In a population of 40,000 individuals, how many would you expect to be heterozygous carriers of the gene for AT? (Hint: Use $2pq$.) What is the approximate frequency (in percent) of such heterozygotes?

8   In order to prevent pollution of the human gene pool, the supporters of the eugenics movement advocated sterilization of the mentally retarded and others with genetic defects. Phenylketonuria (PKU) causes mental retardation, and in the white population perhaps 1 in 10,000 individuals is born with PKU. Would sterilization of all PKU victims greatly reduce the frequency of PKU over the next three hundred years (ten generations)? Justify your answer.

REFERENCES

AYALA, F. J. 1983. *Genetic Variation and Evolution*. Carolina Biology Reader. Carolina Biological Supply Co., Burlington, N.C.

BODMER, W. F., and L. L. CAVALLI-SFORZA. 1976. *Genetics, Evolution, and Man*. W. H. Freeman and Co. San Francisco.

CAVALLI-SFORZA, L. L. 1969. Genetic drift in an Italian population. *Scientific American* 221(2):30–37.

———. 1974. The genetics of human populations. *Scientific American* 231(3):80–89.

GLASS, B., M. S. SACKS, E. F. JAHN, and C. HESS. 1952. Genetic drift in a religious isolate: an analysis of the causes of variation in blood group and other gene frequencies in a small population. *American Naturalist* 86(828):145–59.

GLASS, H. B. 1953. The genetics of Dunkers. *Scientific American* 189(2):76–81.

GLASS, B., and C. C. LI. 1953. The dynamics of racial intermixture—an analysis based on the American Negro. *American Journal of Human Genetics* 5:120.

HARDY, G. H. 1908. Mendelian proportions in a mixed population. *Science* 28:49–50.

HARTL, D. L. 1981. *A Primer of Population Genetics.* Sinauer Associates, Inc. Sunderland, Massachusetts.

METTLER, L. E., and T. G. GREGG. 1969. *Population Genetics and Evolution.* Prentice-Hall, Inc. Englewood Cliffs, New Jersey.

sixteen

Human Genetics for the Citizen

THE GENERAL POPULACE IS INCREASINGLY aware of advances in human genetics and of the importance of genetic diseases to human well-being. This awareness has been brought about by a number of developments. Articles in *Time* and *Newsweek*, television presentations on PBS and on the major commercial networks, and frequent items in newspapers all have contributed to the public awareness of developments in human genetics.

In addition, the educational community is contributing to this awareness: college courses in human genetics for nonscience majors and workshops on human genetics for healthcare professionals and for high school biology teachers are increasingly common. The Center for Education in Human and Medical Genetics, affiliated with the Biological Sciences Curriculum Study, has developed and is testing human genetics educational materials for use with elementary, junior high, and high school students. A federally funded National Clearinghouse for Human Genetic Diseases functions to disseminate genetics educational materials produced by a plethora of governmental and private agencies. The March of Dimes Birth Defects Foundation continues to be an effective producer and distributor of educational materials related to human genetics and birth defects. (See Table 16–1 for addresses of agencies mentioned here.) One genetic counselor—educator in the state of Indiana has made hundreds of presentations on human genetics and birth defects to thousands of high school students in Indiana and other states in the eastern half of the nation. Many other examples could be cited to illustrate current educational developments that are designed to increase genetic literacy in the general public.

277

TABLE 16–1. Three sources of genetics educational materials

(1) The Center for Education in Human and Medical Genetics
c/o BSCS, The Colorado College
Colorado Springs, Colorado 80903

(2) The March of Dimes Birth Defects Foundation
1275 Mamaroneck Avenue
White Plains, New York 10605

(3) National Center for Education in Maternal and Child Health
3520 Prospect Street, NW
Washington, D.C. 20007

Increasing Demand for Genetic Services

Advances in procedures for detecting and treating genetic diseases and increasing public awareness of these advances can be expected to produce a marked increase in the demand for medical genetic services. Increased demand is already in evidence in connection with prenatal detection of genetic defects via amniocentesis; laboratories that handled five or ten amniocenteses per year in the early 1970s have experienced an exponential increase in their caseloads so that by the end of the decade some were dealing with as many as twelve hundred to fifteen hundred cases per year. Although physicians and medical geneticists have served as genetic counselors for years, the increasing demand for medical genetic services and the increasing complexity of the cases encountered have created opportunities for new types of professionals in genetic counseling centers. For example, genetic counseling teams in large hospitals and medical centers may include a theologian, philosopher, or ethicist who can assist both counselors and counselees in better understanding and dealing with the moral, ethical, and theological problems arising in connection with many cases. In addition, genetic counseling teams may include a "genetic counseling associate," a person prepared at the master's degree level to function under the supervision of an M.D. or Ph.D. counselor. Genetic counseling associates are prepared for their positions at a number of colleges and universities around the country; probably the best-known program designed for this purpose is located at Sarah Lawrence College in New York. (See article by J. Marks, cited in References.)

Obtaining Genetic Services

How can the average citizen—the reader of this book—obtain genetic counseling services? One might begin with one's **family physician**, keeping in mind that physicians are not always well schooled in human genetics. (Barton Childs and his colleagues [see References] have documented the rather fragmentary preparation in genetics that physicians receive in U.S. medical schools.) Secondly, the individual or the individual's family physician might contact a **medical school** in their state, since genetic counseling and research units are usually associated with medical schools and major hospitals. Thirdly, the **state medical association** could be expected to provide information on the availability of genetic counseling in the state. Finally, at least two **directories of genetic counseling services** provide information for the entire nation (see References). Arranged by states, these directories provide the names, addresses, and telephone numbers of qualified individuals who direct genetic counseling units.

Genetic Counseling

Genetic counseling was briefly discussed in Chapter 3 of this book. Perhaps at this point more needs to be said about this increasingly important aspect of medicine.

Functions of the Counselor

After seeing a 45-minute-long videotape of a genetic counseling session, a student asked the counselor who had conducted the session, "How do you manage to keep your own sanity when you must deal with such emotionally charged situations on a day-to-day basis?" The counselor quickly replied, "I cry a lot!" He went on to explain that he did not mean to be facetious, but that he felt that it was imperative for the counselor to be able to empathize with the counselee. At the same time the counselor cannot become so emotionally involved that objectivity in viewing medical and scientific data is lost.

These comments emphasize the fact that a genetic counselor must have a number of competencies. (1) A firm grounding in medicine and genetics is without question the foundation upon which effective counseling rests. The counselor must keep abreast of the rapid developments in human and medical genetics in order to provide the best possible service to counselees. (2) The counselor must have the personality traits and the professional counseling skills that enable him or her to work effectively with patients. (3) A genetic counselor must be an effective teacher with the ability and patience to explain, at an intellectual level appropriate to the counselees, what may be rather difficult concepts having serious emotional overtones.

Genetic counselors, no doubt, prefer doing **prospective** genetic counseling. That is, they prefer dealing with counselees who have not yet had a defective child, although they are (or may be) at risk. In such situations the counselor can often help a couple explore alternatives so as to avoid the pain and unhappiness of having a seriously defective child. In fact, however, much genetic counseling is **retrospective**. In these cases the counselor may work with families that already have had a genetically defective child (or children). Courses of action related to the following questions are then investigated: (1) What can be done so that the defective child can achieve its maximum potential? (2) What is the couple's risk of having other children with the defect? (3) What courses of action can prevent such a recurrence? In other cases of retrospective counseling, the counselor works with a counselee who is expecting a child who has a significant risk of being defective (e.g., a pregnant woman who is a "normal" carrier of a 14/21 chromosome translocation and is at high risk for having a Down syndrome child). In such cases the following issues may be explored: (1) Should an amniocentesis for prenatal detection of the defect be done? (2) What are the risks to mother and fetus of the amniocentesis? (3) If the fetus is shown to be defective, should it be aborted?

Genetic counseling demands that (1) an accurate medical **diagnosis** of the condition in question be completed, (2) an accurate and complete family **pedigree** be developed, (3) the **risk factor** for having affected children be calculated and explained, and (4) **alternative courses of action** (options) be explored with the couple. Many genetic counselors take the position that they will be **nonprescriptive** in their counseling, but will gather and explain the necessary data, explore the alternative courses of action available, help prepare the counselees emotionally for these alternatives, and then let the couple decide upon the course of action to be followed. The nonprescriptive counselor takes the position that only the counselees can make reproductive decisions suitable for themselves. Thus, the counselor, in so far as possible, should not let his or her personal values influence that decision.

Other genetic counselors are **prescriptive** in their counseling in that they have in mind the achievement of specific behavior on the part of the counselees. In most situations where we seek the advice of physicians, a prescription or recommended course of action is suggested: "Have this prescription filled at your pharmacy." "Follow this diet." "Plan to have this surgery." The prescriptive genetic counselor thus attempts to lead the counselee to select a particular course of action. Finally, regardless of the philosophical differences in the counseling approach and the option acted upon, the counselees must continue to receive **advice** and **support**.

Genetic Counseling Problems

What are some of the complications encountered in the genetic counseling process? **Obtaining a correct diagnosis** of the disease or defect is, of course, imperative. Because of complications created by incomplete penetrance, by phenocopies, and by phenotypically similar conditions having different genetic or chromosomal etiologies, obtaining an accurate diagnosis is often not easy to do. The services of a variety of medical specialists and the use of various laboratory tests may be necessary for correct diagnosis.

Developing a complete and accurate family history or pedigree may also prove time-consuming and difficult. Since key individuals in the family may be living at great distances from the patient and counselor, provisions may have to be made for bringing these individuals to the counselor or for a medical team to go to the individuals. Other members of the family may be deceased and the counseling team may be forced to rely on medical records of the deceased for any accurate information they can get. Related to developing a pedigree is the problem of **confidentiality of medical records.** What can be done if a relative does not want to share medical information that might be helpful to the counselee? Ought potentially affected relatives of the counselee be informed of their own risk? What are the consequences of informing (or not informing) them? What are the responsibilities of the counselor with respect to telling the counselee the whole truth—especially when the truth may shatter a marriage or otherwise seriously disrupt the counselee's life? These are all thorny medical-ethical-legal issues that the counselor and often the counselee may face. Their discussion is beyond the scope of this book.

Certainly **explaining risk factors**—getting the counselees to understand the meaning of probability—is a difficulty encountered in many cases. Counselors may resort to a variety of teaching gimmicks to help the counselee grasp the meaning of probability. Indeed, difficulties in understanding risk factors may be symptomatic of the overall difficulty the counselees have in understanding the genetic problem with which they are confronted. One counselor said that he explains the findings to the counselees, giving them frequent opportunities to raise questions. He then asks them appropriate questions to determine whether they understand what they have been told. After some time has elapsed, perhaps at the next counseling session, a second counselor meets with the couple and again explains the findings and the options open to the couple. This approach appears to be modeled on an experience many of us have had: sometimes one teacher can effectively communicate with a student and enable the student to understand a difficult concept, despite the fact that another teacher was unsuccessful in doing so. Finally, the findings are communicated in writing to the counselees. This approach gives the counselees the opportunity to restudy the findings and the options open to them, and it also serves to protect the counselor, documenting that the information has been correctly communicated to the counselees.

Many cases encountered by the genetic counselor are filled with serious **emotional overtones** that must be treated in counseling sessions. The parents whose newborn has been diagnosed as having Down syndrome, the young Jewish couple whose six-month-old, firstborn son has been diagnosed as having Tay-Sachs disease, or the couple with a severely retarded microcephalic child can reasonably be expected to have feelings of disbelief, anger, guilt, and depression. It has been noted that the parents of a seriously defective child often experience the same psychological patterns observed in the dying patient who is told that he has at most six months to live. The genetic counselor must be prepared to deal with such emotional reactions and with family disputes in which one parent blames the other for the child's defect; and the counselor, as the bearer of unpleasant news, must be able to accept personal rejection.

Advances in Genetics

In most cases in the past the genetic counselor could only deal with recurrence risk data. If a couple had had a child with sickle-cell anemia or with cystic fibrosis, they could be told that there was a one-in-four risk of recurrence for each pregnancy. Today, however, the genetic counselor often has other options in dealing with counselees. **Genetic screening** procedures can be used to determine whether individuals in certain high risk populations are carriers of "recessive" genes for such diseases as sickle-cell anemia, Tay-Sachs, galactosemia, and (more recently) cystic fibrosis. If only one member of the couple being counseled is heterozygous for one of these diseases, and the other member is homozygous for the "normal" allele, then there is *no* risk for having an affected child. If both husband and wife are found to be heterozygous, of course, the risk is one in four. Thus, by being able to identify heterozygotes with certainty, more complete and accurate prospective genetic counseling can be done.

On the other hand, suppose that both the husband and wife are found to be carriers of the Tay-Sachs gene and the wife is already pregnant. In this case the option of **amniocentesis** makes possible prenatal detection of the genetic condition of the fetus. If the fetus is found to be homozygous recessive for the Tay-Sachs allele, the couple then has the option of terminating the pregnancy. As mentioned in Chapter 3, prenatal detection of 182 conditions has been documented. Although prenatal detection increases the options available to the counseled couple, it also may create moral-ethical problems for them if they must consider terminating a pregnancy. Here again, the counselor must be sensitive to the counselees' concerns and supportive of whatever decision they reach.

Other alternatives now available also create moral dilemmas for the counselees. If both husband and wife are heterozygous for a recessive deleterious allele, still want to have children, but do not want to take the one-in-four risk of having a defective child, they could resort to **artificial insemination** using **donor** semen (AID). The heterozygous woman could be inseminated artificially using semen from a man known to be homozygous for the normal allele. Although assuring the birth of a child that is not homozygous recessive for the deleterious gene in question, AID may be seriously questioned by the couple on moral, ethical, or theological grounds. Again we have an illustration of a bioethical problem arising in the context of genetic counseling. One can readily understand why a genetic counseling team may include a theologian, philosopher, or ethicist.

A couple at risk may not want to have children at all, and may want professional advice on **contraception.** The genetic counselor should be prepared to discuss alternate means of contraception and the reliability and potential dangers of each. Moral and ethical issues related to contraception may also require special counseling for some couples.

Genetic Counseling and Eugenics

As implied in Chapter 13, the **eugenics movement** was concerned with reducing the frequency of harmful or deleterious genes in the human population. Genetic counselors, with their improved armamentarium of weapons, are concerned with helping individuals and families afflicted by deleterious genes. Secondarily, desirable eugenic effects can be considered, but the primary obligation of genetic counselors (in the opinion of most medical geneticists) is to their clients or counselees, and not to society in general. On the other hand, it would not appear to be unreasonable for genetic counselors to point out the implications for society of their counselees' reproductive decisions. The counselees may then wish to give serious consideration to long-term societal consequences of their personal reproductive decisions.

Where Do You Go From Here?

In this book you have had an opportunity to explore some of the basic principles of genetics as they apply to human beings. Perhaps you still have questions that were not answered, and you will want to seek additional information in more advanced and technical references. Some of you may even have questions for which medical geneticists have no answers at the present time. This would not be surprising. Human and medical genetics has made tremendous progress in the last quarter of a century, but it is still an emerging science with new and significant discoveries being made almost daily. Probably all of you have thought of questions pertaining to inheritance in your own families and would like to obtain answers to these questions.

If we have caused you to raise any of these types of questions—if we have stimulated you to want to learn more about human genetics, then perhaps we should be satisfied that we have achieved our goal in writing this book. Perhaps we are overly idealistic, but we might hope for more: Being aware of genetic diseases and defects, perhaps you will see the need to encourage and support research in medical genetics. Perhaps you will see fit to encourage your federal and state legislators to provide funds for genetic services, research, and education. Perhaps you will want to contribute some of your time, talent, and financial resources to one of the many private agencies concerned with specific genetic diseases (see Table 16–2 for selected examples). Perhaps, out of concern for your own family or for society in general, you will seek genetic counseling if you suspect there is need for it.

In any event, we hope that you finish this book not only better informed about the principles of human genetics, but also with an expanded consciousness about the significance of human genetics for human well-being. Only if we have a genetically literate and conscious society can we expect the necessary resources to be made available to understand, prevent, and control genetic diseases and defects. If we have convinced you of the significance of human genetics for human welfare, we hope you will act on your convictions.

TABLE 16–2.   Selected organizations concerned with research, treatment, and education relative to specific genetic diseases

(1) Cooley's Anemia Blood and Research Foundation
Graybar Building, Suite 1644
420 Lexington Avenue
New York, New York 10017

(2) Cystic Fibrosis Foundation
3379 Peachtree Road, N.E.
Atlanta, Georgia 30326

(3) National Hemophilia Foundation
19 West 34th Street, Suite 1204
New York, New York 10001

(4) Committee to Combat Huntington's Disease, Inc.
250 West 57th Street, Suite 2016
New York, New York 10107

(5) Muscular Dystrophy Association
810 Seventh Avenue
New York, New York 10019

(6) National Association of Sickle Cell Disease, Inc.
945 South Western Avenue
Los Angeles, California 90006

(7) Spina Bifida Association of America
343 South Dearborn, Room 319
Chicago, Illinois 60604

(8) National Tay-Sachs and Allied Diseases Association
122 East 42nd Street
New York, New York 10017

PROBLEMS AND QUESTIONS

1 Match the correct lettered terms with their numbered definitions or descriptions.

definitions or descriptions

(1) Genetic counseling of a couple who is at risk, but who have not yet had a defective child, may be spoken of as ____ counseling.

(2) Genetic counseling in which the counselor explores alternatives but encourages the counselees to choose the option with which they feel most comfortable, is spoken of as ____ counseling.

(3) A possible source of complication in attempting to determine the mechanism of inheritance of a particular trait.

terms

(a) prescriptive
(b) amniocentesis
(c) incomplete penetrance
(d) retrospective
(e) risk factor
(f) nonprescriptive
(g) contraception
(h) prospective

2 List several possible sources of information about genetic counseling services.

3 List several reasons why genetic counseling teams may include an ethicist, philosopher, or theologian.

4 List several recent developments in medical genetics that could create moral-ethical problems for counselees.

5 If you were to select a genetic counselor, what personal and professional characteristics would you want that counselor to possess?

REFERENCES

CHILDS, B., C. A. HUETHER, and E. A. MURPHY. 1981. Human genetics teaching in U.S. medical schools. *American Journal of Human Genetics* 33(1):1–10.

CLINICAL GENETIC SERVICE CENTERS, A NATIONAL LISTING. 1980. U.S. Dept. of Health and Human Services. DHHS Publication No. (HSA) 80–5135.

FUHRMANN, W., and F. VOGEL. 1983. *Genetic Counseling*, 3d ed. Springer-Verlag. New York.

LYNCH, H. T., P. FAIN, and K. MARRERO. 1980. *International Directory of Genetic Services,* 6th ed. March of Dimes. White Plains, New York.

MARKS, J. H. 1979. Masters level training programs for genetic counselors: an eight year report. In *Service and Education in Medical Genetics* edited by I. H. Porter and E. B. Hook. Academic Press. New York.

REED, S. 1980. *Counseling in Medical Genetics,* 3d ed. A. R. Liss, Inc. New York.

RHINE, S. A. 1983. The most important nine months. *The Science Teacher* 50(7):46–51.

acentric A chromosome without a centromere.

acrocentric A chromosome with the centromere attachment near one end, making one arm very short.

adenine One of the nitrogenous bases found in the nucleotide building blocks of both DNA and RNA; a purine.

allele (short for **allelemorph**) One of two or more alternative forms of a gene at a particular locus on a chromosome.

amniocentesis Removal of some amniotic fluid from around the fetus by means of a slender needle inserted through the abdomen of the mother. Used to obtain fetal cells and amniotic fluid, which will be studied to identify abnormalities of the fetus.

amnion A membrane surrounding the fetus. Contains amniotic fluid in which the fetus floats.

aneuploid An organism whose chromosome number is not an exact multiple of the haploid number of chromosomes; e.g., a trisomic or monosomic.

antibody A plasma protein, gamma globulin, or immunoglobulin, produced in response to contact with a foreign antigen. Will react with and neutralize a specific foreign antigen.

antigen A substance that can stimulate the production of specific antibodies.

autosome Any chromosome that is not a sex chromosome (not an X or Y chromosome).

Barr body (sex chromatin body) A body that lies against the inside of the nuclear membrane of normal female cells and that stains heavily with basic stains. It is a tightly coiled X chromosome.

carcinogen An agent that induces cancer.

centromere A constriction on a chromosome that holds together the two chromatids of a prophase chromosome; the site on the chromosome where the spindle fiber attaches during mitosis and meiosis.

chorion A highly vascular outer embryonic membrane that enters into the formation of the placenta.

chromatid One of two identical halves of a chromosome after it has duplicated, but before separation to form separate chromosomes. Chromatids show clearly in the prophase; each chromosome is made of two chromatids that are joined by a single centromere.

chromosome Gene-carrying body that is threadlike in interphase and rodlike during mitosis. Contains protein (histone) as well as the DNA of the genes.

clone A group of cells derived from a single cell by repeated mitosis; the cells of a clone are genetically identical.

codominant Full expression of two allelic genes in a heterozygous person. The alleles for the A and B blood antigens are codominant.

codon A triplet of bases in messenger-RNA that codes for a specific amino acid in the formation of a polypeptide chain.

complementation test A procedure for determining whether two mutant genes are alleles of each other; if both mutant genes are present in a cell line and the phenotype is wild type, then the mutant genes are not alleles of each other.

concordance   The degree of similarity between twins or sibs according to specific traits. Usually expressed as a percentage.

consanguineous   Refers to marriages, or matings, of genetically related persons.

cytosine   One of the nitrogenous bases found in the nucleotide building blocks of both DNA and RNA; a pyrimidine.

deletion   The loss of a portion of a chromosome.

dichorionic   Term used to describe twin embryos that have separate chorions.

diploid   Having two of each kind of chromosome ($2n$). Normal somatic cells are diploid since each chromosome in a cell has a mate of the same kind. (Exception: X and Y in males are different.)

dizygotic   Refers to twins who originate from two different zygotes (fertilized eggs). Also known as fraternal twins.

DNA (deoxyribonucleic acid)   The macromolecule of which genes are made.

dominant   Refers to characteristics that are expressed when a person is either homozygous or heterozygous for a certain gene. Also used to refer to genes that are expressed in spite of the presence of recessive alleles.

drift, genetic   Change in the frequency of genes in a population as a result of chance fluctuations. Often comes about when a small isolate, with genes not representative of its population, migrates to a new region.

egg   An unfertilized ovum or female gamete.

embryo   An organism in early developmental stages. In humans, *embryo* generally refers to the developing individual up to the beginning of the third month of pregnancy.

enzyme   A protein that catalyzes a specific chemical reaction.

epistasis   A case in which a gene at one locus suppresses the action of another gene (or genes) at a different locus. The gene for albinism is epistatic to nonallelic genes that govern the intensity of melanin deposits.

euchromatin   The part of a chromosome that is rich in DNA (genes), in contrast to heterochromatin. Euchromatin is condensed during mitosis but uncoiled during interphase.

eugenics   The improvement of the human population by altering its genetic composition.

fetus   The human embryo after it has assumed the form of a human body, typically from the latter part of the third month of development until birth. (See also **embryo**).

founder principle   Changes in gene frequency occur in small populations because of their origin from only a few individuals (the founders) separated from the original population.

gamete   A reproductive cell. A sperm or an egg.

gene   The basic unit of heredity. Occupies a fixed locus on a chromosome. A unit of DNA that codes the production of a single polypeptide chain.

gene frequency   The number of gene loci (in a particular population) at which a specific allele is present divided by the total number of loci at which the allele could occur.

gene pool   The total of all the genes in a population.

gene splicing   Any technology used to combine DNA from different organisms; i.e., recombinant DNA technology. Characteristically, enzymes of various types are used to "cut" and then "splice" DNA segments.

genetic counseling   Advising those at risk for having a genetically or chromosomally defective child of those risks; counseling parents who have produced a genetically defective child.

genetic death   Failure to transmit genes to future generations. Can result from death before maturity or from failure to reproduce after maturity.

genetic engineering   Any human intervention designed to alter, at the molecular level, the genetic constitution of an organism. See also **gene splicing.**

genotype   The type of genes in an individual; the genetic constitution of an individual. Compare **phenotype.**

germ plasm   The reproductive cells and their immediate precursors; also, the genetic material transmitted through the germ cells.

globulin   A class of proteins soluble in dilute salt water, but insoluble in pure water. The human blood plasma contains alpha, beta, and gamma globulins.

guanine   One of the nitrogenous bases found in the nucleotide building blocks of DNA and RNA; a purine.

haploid   A single set of chromosomes, one of each kind, the *n* number. Sperm and eggs are haploid, in contrast to other body cells, which are diploid, that is, have two of each kind of chromosome.

haptoglobin   An alpha globulin in the blood plasma that binds and carries free hemoglobin from broken red blood cells.

Hardy-Weinberg law   The basic principle of population genetics, which holds that gene frequencies and genotype frequencies remain constant generation after generation in large random-mating populations where there is no mutation, selection, migration, or genetic drift.

hemizygous   The haploid state of a gene or genes in an otherwise diploid individual. Males are hemizygous for most of the genes on their single X chromosome.

hemoglobin   Iron-containing protein in red blood cells. Transports oxygen.

hermaphrodite   Individual with both male and female reproductive organs. Has somewhat intermediate secondary sex characteristics.

heterochromatin   The part of a chromosome that is low in DNA concentration, in contrast to euchromatin. Shows maximum staining at interphase; fluoresces with the Q-banding staining technique.

heterozygous   Having different alleles for a particular gene locus.

holandric   A gene on the Y chromosome and therefore expressing only in males.

homozygous   Having the same alleles for a particular gene locus.

immunogenetics   The study of the role of genetics in the immune reaction. How heredity affects antibody production and function.

immunoglobulin   Blood plasma globulin that forms antibodies. A gamma globulin.

implantation Attachment of an embryo to, and implantation within, the wall of the uterus.

independent assortment The chance distribution to the gametes of genes located on separate chromosome pairs. Mendel's second law.

intermediate inheritance The expression of both genes of an allelic pair to an approximately equal degree, resulting in a phenotype about halfway between that of the two homozygotes. Sometimes called lack of dominance.

karyotype A paste-up of the chromosomes from a photograph of a cell. The chromosomes are arranged in matched pairs according to length, centromere position, and banding pattern.

kinetochore Synonym of **centromere.**

lethal gene A gene that causes death when it is expressed. In many cases death occurs before or shortly after birth.

linked genes Genes that are located on the same chromosome.

locus (plural, **loci**) The position on a chromosome occupied by a certain gene. All allelic genes have the same chromosome locus.

meiosis A series of two cell divisions accompanied by only one gene and chromosome replication. Reduces the chromosome number from the diploid to the haploid; occurs in animals when gametes are produced.

messenger RNA (mRNA) The RNA transcribed from DNA and coding for the amino acid sequence of a polypeptide.

metacentric A chromosome with the centromere near the middle so both arms are of about the same length.

mitosis The series of reactions that brings about the duplication and segregation of chromosomes so that the two daughter cells formed each have the same number and kind of chromosomes as were present in the original cell.

monochorionic Term used to describe twin embryos that share the same chorion.

monosomy The condition in which one chromosome of a given pair is missing; i.e., $2n - 1$. A person with monosomy-21 would have only one chromosome 21 in the body cells.

monozygotic Refers to twins who originate from a single zygote (fertilized egg). Also known as identical twins.

mosaic, genetic A person with two or more cell lines differing in genes or chromosomes, usually the result of mutation or chromosome aberration in the early embryo.

mutagen A physical or chemical agent that induces mutation; e.g., X rays, nitrogen mustard.

mutation Change in the DNA sequence of a gene that causes it to code a different kind of protein. May result in an alteration of the phenotype when expressed. (Sometimes used to refer to chromosome aberrations as well.)

nondisjunction The failure of chromosomes to disjoin (separate) at metaphase of meiosis or mitosis. Both chromosomes of a kind thus go to one cell, while the other cell gets no chromosome of this kind.

nucleic acid   A chain of many nucleotides; each nucleotide consisting of a pentose sugar, a nitrogenous base, and a phosphate. May be DNA or RNA.

oocyte   Cell in the ovary that forms an egg and polar bodies through meiosis. Primary oocyte undergoes meiosis I to form secondary oocyte and a polar body; secondary oocyte undergoes meiosis II to form an egg and a polar body.

oogenesis   Process of egg formation through the two divisions of meiosis.

pedigree   A chart or diagram showing the ancestral history of a person for a particular trait or for a number of traits.

penetrance   The proportion of persons who express certain genes they carry. A dominant gene with complete penetrance is expressed by all who carry it. A dominant gene with 80 percent penetrance is not expressed by 20 percent of the carriers. This is called *incomplete penetrance.* Homozygous recessives may also have incomplete penetrance.

phenocopy   A trait induced by environment that is a duplication of a trait caused by genes in other instances.

phenotype   The appearance of the individual; phenotype results from the interaction of the individual's genotype with the environment

placenta   Consisting of both uterine (maternal) and embryonic tissue, this organ permits the developing embryo to be nourished and to eliminate wastes.

pleiotropy   Multiple phenotypic effects of a single gene; the effects may appear to be distinct and unrelated.

polar body   Small cell produced from the primary and secondary oocytes as a result of the unequal division of these cells during meiosis.

polygenic   Genes at a number of different loci influencing the expression of a single phenotypic trait, such as stature.

polymorphism   The occurrence of two or more genetically different types of individuals within the same interbreeding population (e.g., individuals with blood types A, B, AB, and O).

polypeptide chain   Chain of amino acids assembled at the ribosomes according to the code carried by messenger-RNA. May be a protein molecule or may combine with other chains to form a protein molecule.

polyploid   An individual having more than two sets of chromosomes; e.g., a triploid ($3n$) or a tetraploid ($4n$).

population genetics   The study of the genetic composition of groups of organisms or populations. See **Hardy-Weinberg law.**

pseudohermaphrodite   Individual possessing the gonads of one sex, but expressing sexual characteristics of both sexes.

recessive   Refers to characteristics that are expressed only when an individual is homozygous or hemizygous for the gene that specifies them. Also used to refer to the genes that are not expressed when a dominant allele is present.

replication   A duplication. Refers to genes, chromosomes, and cells. Replication of DNA requires copying from a template DNA.

ribosomes   Small bodies in the cytoplasm of cells. The site of protein synthesis. Made of protein and ribosomal RNA (rRNA).

RNA Ribonucleic acid, a single-stranded nucleic acid, usually transcribed from DNA. Messenger RNA, transfer RNA, and ribosomal RNA are three forms of this nucleic acid.

satellite A small part of an acrocentric chromosome attached to the short arm of the chromosome by a slender stalk.

screening, genetic Any one of many procedures used to identify individuals afflicted with a particular genetic disease or who carry in the heterozygous condition a "recessive" allele for a genetic disease.

segregation Refers to the separation of genes of a pair. Segregation results from chromosomal separation in meiosis I during gamete formation. Mendel's first law.

sex chromatin body Synonym for **Barr body.**

sex chromosomes Chromosomes involved in sex determination; the X and Y chromosomes.

sex influenced Describes genes or traits that vary in their dominance in the two sexes. A gene may be dominant in males and recessive in females, or vice versa.

sex limited A characteristic expressed in one sex only; could be due to autosomal genes or genes on the sex chromosomes.

sex linked See **X linked.**

sibs, siblings Brothers and/or sisters born of the same parents.

somatic mutation Any mutation that occurs in body cells other than those that develop into germ cells.

somatoplasm All of the body except the reproductive cells and their immediate precursors. Opposite of **germ plasm.**

sperm A male gamete.

spermatid Cell formed by spermatogenesis. Becomes a sperm.

spermatocyte Cell in the testes that can form sperm through meiosis. Primary spermatocyte is a cell ready for meiosis; divides and forms two secondary spermatocytes, which divide to form four spermatids.

spindle The cytoplasmic structure formed during mitosis and meiosis on which chromosomes align prior to being separated to opposite poles.

synapsis The pairing of homologous chromosomes, seen in the prophase of meiosis I.

syndrome A series of abnormalities arising from a single cause. Many individual genes result in syndromes, or a chromosome aberration may cause a syndrome.

teratogenic Refers to any agent that causes abnormalities in an embryo when a pregnant women is exposed.

tetraploid Having four of each kind of chromosome ($4n$).

thymine One of the nitrogenous bases found in the nucleotide building blocks of DNA; a pyrimidine. Thymine does not occur in RNA.

transduction The transfer of genetic material (DNA) from one organism to another by means of a viral vector.

transferrin Beta globulin of blood plasma that transports iron.

transfer RNA (tRNA) An RNA molecule that carries (transfers) an amino acid to a developing polypeptide chain during the course of translation.

transformation The genetic modification of one strain of organism by the "transforming substance" (DNA) of another strain.

translocation A portion of a chromosome that has broken off and becomes attached to another chromosome. Also refers to an entire chromosome that has become attached to another chromosome.

triploid Having three of each kind of chromosome ($3n$).

trisomic Having three of one kind of chromosome in an otherwise normal diploid cell ($2n + 1$). Trisomy-21 indicates that three 21s are present.

uracil One of the nitrogenous bases found in the nucleotide building blocks of RNA; a pyrimidine. Uracil does not occur in DNA.

X chromosome One of the sex chromosomes associated with sex determination. Normal females have two X chromosomes; normal males have only one.

X linked Refers to genes on the X chromosome that have no alleles on the Y chromosome. Males express all the X-linked genes they carry.

Y chromosome One of the sex chromosomes that triggers the expression of male characteristics. Present in males, but not in females as a rule.

Y linked Refers to genes on the Y chromosome that have no alleles on the X chromosome. Males express all their Y-linked genes. See also **holandric.**

zygote A cell produced by the union of a sperm and an egg. A fertilized egg.

Answers—Chapter One

1 If 12 percent of the bases are *A*, then 12 percent should also be *T* since *A* and *T* are always paired with each other. Thus *A* and *T* account for a total of 24 percent of the bases, leaving 76 percent to be equally divided between *C* and *G*; i.e., 38 percent *C* and 38 percent *G*. DNA contains no uracil (*U*), a base characteristic of RNA.

2 DNA————→RNA————→Protein————→Characteristic

3 The nucleotide building blocks in RNA will have ribose sugar (*R*) instead of deoxyribose (*D*), and uracil (*U*) will replace thymine (*T*). Thus the four building blocks can be symbolized

$$\text{as } \overset{R-A}{p},\ \overset{R-U}{p},\ \overset{R-C}{p},\ \text{and } \overset{R-G}{p}$$

4 GUA CAC UUA ACA CCA GAG GAG AAA

5 The difference between normal and sickle-cell hemoglobin involves the sixth group of three bases, which read CTC in the gene for normal hemoglobin but CAC in the gene for sickle-cell hemoglobin.

6 The change from CTC in the normal gene to CAC in the sickle-cell gene can be expected to alter one amino acid in the amino acid sequence in the hemoglobin. This is, in fact, what happens with the sixth amino acid, glutamic acid, being replaced by valine.

7 A mutation might be defined as the substitution of one base for another in a DNA code for a sequence of amino acids. This substitution of the "wrong" base results in placing the "wrong" amino acid in the protein that the gene regulates, and this results in a change in one of the characteristics of the organism.

8 bases, DNA, chromosome, nucleus, cell, organism

9 To most biologists, the striking similarity between human and chimpanzee chromosomes is regarded as evidence of a relatively close relationship between the two species. In terms of evolution, most biologists would suggest that humans and chimpanzees are modern descendents of a common ancestor.

Answers—Chapter Two

1 (a) 78; (b) 39; (c) 39; (d) 39; (e) 78; (f) 39

2 G_1, S, G_2, prophase, metaphase, anaphase, telophase

3 (a) 7,932; (b) 3,966; (c) 7,932

4 (a) 1,983; (b) 1,983; (c) 5,949

5 46 chromosomes; 92 chromatids; 23 tetrads

6 (a) true; (b) false (Meiosis does consist of two cycles of cell division, but the genes are duplicated only before the first cycle); (c) true; (d) true

7 **(a)** metaphase of meiosis I; **(b)** metaphase or anaphase of meiosis II; **(c)** prophase of meiosis I.

8 **(a)** primary spermatocyte; **(b)** spermatids; **(c)** secondary spermatocyte; **(d)** sperm

9 **(a)** secondary oocyte; **(b)** primary oocyte; **(c)** polar bodies

Answers—Chapter Three

1 The child developing from an unfertilized human egg, the chromosomes of which have doubled, should be a female. Such an individual would be expected to be a female having two X chromosomes resulting from the doubling of the one X chromosome of the egg. To develop into a male, a human fertilized egg must possess one X and one Y chromosome. The latter could only be obtained from a sperm by fertilization.

2 The fact that the two identical twins do not exhibit the same behavior patterns with respect to tongue rolling and hand folding suggests that these may be learned rather than inherited behaviors.

3 fertilized egg, two-cell stage, morula, blastocyst, fetus, newborn

4 For a listing of genetic counseling centers in your area, see the following directories:

U.S. Dept. of Health and Human Services. 1980. *Clinical Genetic Service Centers, A National Listing.* DHHS Pub. No. (HSA) 80–5135. (Available from the National Clearinghouse for Human Genetic Diseases, P.O. Box 28612, 805 15th Street, N.W., Suite 500, Washington, D.C. 20005.)

Lynch, H. T., P. Fain, and K. Marrero. 1980. *International Directory of Genetic Services,* 6th ed. (Available from the March of Dimes Birth Defects Foundation, 1275 Mamaroneck Avenue, White Plains, New York 10605.)

5 *DNA replication,* a process that results in the production of identical DNA molecules (genes) and *mitosis,* a process that produces daughter cells containing identical chromosome sets.

6 X rays, a form of ionizing radiation, are known to be teratogenic, and could result in the production of a deformed fetus. X rays are also mutagenic and could produce genetic change in either the fetus or the mother. The sound waves used in making a sonogram are not harmful to the fetus or the expectant mother.

Answers—Chapter Four

1

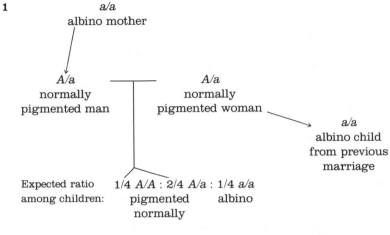

Expected ratio   1/4 A/A : 2/4 A/a : 1/4 a/a
among children: pigmented   albino
                  normally

2   *C/c* × *C/c*   (parents of normal woman)

1/4 *C/C* : 2/4 *C/c* : 1/4 *c/c*

We know that the parents of the normal (non-CF) woman must have been heterozygous carriers of the CF gene since they also had a daughter (*c/c*) who died of CF at age two years. Among the normal children of such a marriage 1/3 will be *C/C* and 2/3 *C/c*. Thus, the normal woman has a 2/3 probability of being heterozygous. If she marries a *C/C* man she would not be expected to have any CF children even if she is a carrier. If she is a carrier and marries a carrier male, the probability is 1/4 that she will have a CF child.

3   Most CF children (genotype *c/c*) are the offspring of marriages in which both parents are *C/c* (phenotypically normal carriers); i.e., CF is caused by a recessive allele. Huntington's disease, by contrast, is due to a dominant gene. This means that a person who inherits the gene from one parent only will develop the disease. The parent who transmits the gene, of course, will also have the disease; i.e., most marriages where Huntington's disease occurs involve one "normal" parent (*h/h*) and one affected parent (*H/h*). The probability is thus 50 percent that the child will inherit the *H* gene and develop the disease.

4   People who have sickle-cell anemia have a disease that is often debilitating; they are homozygous (*Hb^S/Hb^S*) for the sickle-cell gene. People who have sickle-cell trait are generally free of symptoms and are heterozygous (*Hb^A/Hb^S*). They produce both normal and sickle hemoglobin.

5   **(a)**   four—*AB, Ab, aB,* and *ab*
     (b)   1/4 for *A/a B/b*; 1/4 for *a/a B/b*; 0 for *A/A b/b*
     (c)   9/16 for *A–B–*; 3/16 for *A–b/b*; 1/16 for *a/a b/b*

6 (a) A/A × A/A ⟶ all A/A

        normal × normal ⟶ all normal

                     or

        A/A × A/a ⟶ 1/2 A/A + 1/2 A/a

        normal × normal ⟶ all normal

  (b) A/a × A/a ⟶ 3/4 $A-$: 1/4 a/a

        normal × normal ⟶ 3 normal : 1 albino

  (c) A/A × a/a ⟶ all A/a

        normal × albino ⟶ all normal

  (d) A/a × a/a ⟶ 1/2 A/a + 1/2 a/a

        normal × albino ⟶ 1 normal : 1 albino

  (e) a/a × a/a ⟶ all a/a

        albino × albino ⟶ all albino

  (f) a/a C/C × A/A c/c ⟶ all A/a C/c

        albino × albino ⟶ all normal

This mating is possible because, in fact, two dominant genes (A and C) are necessary for producing melanin pigment. Because a/a C/C is homozygous recessive for a, albinism results. Because A/A c/c is homozygous recessive for c, albinism results. The A/a C/c offspring possess both dominant genes and can produce melanin.

  (g) A/a c/c × a/a C/c ⟶ 1 A/a c/c : 1 a/a C/c : 1 a/a c/c : 1 A/a C/c

        albino × albino ⟶ 3 albino : 1 normal

  (h) A/a C/c × A/a C/c ⎫   9/16 $A-C-$ normal

        normal × normal   ⎬ ⟶   ⎧ 3/16 $A-c/c$ albino

                              7/16 ⎨ 3/16 a/a $C-$ albino

                                   ⎩ 1/16 a/a c/c albino

Answers—Chapter Five

1   sperm

		X	Y	XY	XX	YY	O
	X	XX 1	XY 2	XXY 3	XXX 4	XYY 5	XO 6
eggs	XX	XXX 7	XXY 8	XXXY 9	XXXX 10	XXYY 11	XX 12
	O	XO 13	YO 14	XY 15	XX 16	YY 17	OO 18

The checkerboard shows that eighteen combinations are possible. Note that the union of certain kinds of abnormal gametes can produce a chromosomally normal fetus (see squares 12, 15, and 16). Some combinations completely lack an X chromosome and are lethal (see squares 14, 17, and 18). Squares 4 and 7 represent trisomic-X females and square 10 is a tetrasomic-X female. Squares 6 and 13 represent Turner syndrome (monosomic-X) females. Individuals represented by squares 3, 8, 9, and 11 have variations of Klinefelter syndrome. Individual 5 has Jacobs karyotype.

2 Number of sex chromatin bodies for each individual in the checkerboard in answer 1 above: (1) one; (2) none; (3) one; (4) two; (5) none; (6) none; (7) two; (8) one; (9) two; (10) three; (11) one; (12) one (13) none; (14) none; (15) none; (16) one; (17) none; (18) none

3 Males: $105/205 \times 3,000,000 = 1,536,585$; Females: $100/205 \times 3,000,000 = 1,463,414$

4 The birth of "extra" males may be an inherent compensatory mechanism to insure a 1 : 1 sex ratio by the time the reproductive years are reached. Thus, at twenty years of age, there is one male for every female. If equal numbers of males and females were live-born, by age twenty there would be a deficiency of males because of the differential death rate.

5 **(a)** 44 autosomes + XXXY = 48 total; male
 (b) 44 autosomes + X = 45 total; female
 (c) 44 autosomes + XXYY = 48 total; male
 (d) 44 autosomes + XXXX = 48 total; female

Answers—Chapter Six

1 No, the son did *not* inherit his father's vision defect; the son received a Y chromosome from his father and an X chromosome, which carried the recessive gene for deutan colorblindness, from his mother.

$$X^D X^d \quad \times \quad X^d Y \quad \longrightarrow \quad X^d Y$$

$$\text{normal mother} \qquad \text{colorblind} \qquad\qquad \text{colorblind}$$
$$\text{father} \qquad\qquad\qquad \text{son}$$

2 $\dfrac{d \quad P}{d \quad P} X\!\!-\!\!X$ $\times$ $\dfrac{D \quad p}{\quad} X\!\!-\!\!y$ $\longrightarrow$ $\dfrac{d \quad P}{D \quad p} X\!\!-\!\!X$ and $\dfrac{d \quad P}{\quad} X\!\!-\!\!y$

woman man daughter son
(deutan) (protan) (normal) (deutan)

3 The trait described in this problem seems to be regulated by an X-linked dominant gene. If we let T = the dominant allele for the trait and t = the recessive allele for "normal" (absence of the trait), then the genotypes of the individuals described in the problem could be represented as follows:

$$X^T Y \quad \times \quad X^t X^t \quad \longrightarrow \quad X^T X^t \quad \text{and} \quad X^t Y$$

affected "normal" affected "normal"
man wife daughter son

$$X^T X^t \quad \times \quad X^t Y \quad \longrightarrow \quad X^T X^t \quad \text{and} \quad X^t X^t$$

affected "normal" affected "normal"
daughter husband daughter daughter

$$X^T Y \quad \text{and} \quad X^t Y$$

affected "normal"
son son

4   If we assume pattern baldness to be an example of a sex-influenced trait, then one can postulate the following genotypes where B = the allele for normal hair and B' = the allele for pattern baldness:

BB'	×	BB'	⟶	$B'B'$	and	BB
unaffected woman		pattern bald man		pattern bald daughter		unaffected son

5   (1) (d) and (g); (2) (a) and (h); (3) (c) and (e); (4) (b) and (f)

Answers—Chapter Seven

1   A high resistance to infections is accomplished by a very efficient immune system, but such a system would produce high antibody titers and these might react with body tissues and cause autoimmune diseases.

2   The diseases listed first have long incubation periods so there is time for memory cells to be reactivated before symptoms appear, but this is not true for the common cold; its symptoms appear before the memory cells can produce their antibodies.

3   The fetus might produce antibodies against the mother's tissue at the point of contact with the placenta. This would cause a rejection of this tissue as foreign and would result in early abortion.

4   Immunosuppressant drugs must be given regularly following a kidney transplant to prevent rejection. Since the immune system protects us against cancer to some extent, susceptibility to cancer would increase with the loss of an effective immune system.

5   Yes, in both instances; the genes would be identical, so histocompatibility antigens would match as in identical twins.

6   Different infectious diseases occur in the two regions, so there would be selection favoring different HLA types in the two regions. Different HLA types would cause different autoimmune diseases.

7   They may have been sensitized by leukocytes that gained entrance to their bodies from fetuses during pregnancies.

8   Most autoimmune diseases are the result of sensitization to a particular antigen, but those against virus nucleic acid can be different because these nucleic acids vary. Hence different tissues may be affected.

9   These mutagenic agents might cause mutations in lymphocytes that cause them to change and produce antibodies against one's own tissue.

Answers—Chapter Eight

1   The type A mother must be A/a and the type B father, A^B/a. The children would be a/a (type O), A/a (type A), A^B/a (type B) and A/A^B (type AB).

2   No, the woman and her husband could both be heterozygous and carriers of the a(O) allele; the child would be homozygous a/a (type O).

3 Yes; the child is not their baby. The gene for M would have come from the father so the child would have to be MN and could not be N.

4 As donors, yes, but not as recipients. They would not have any A or B antigens to react with anti-A or anti-B antibodies in the blood of recipients. They would not be suitable recipients, however, because they would have antibodies that would react with both donor blood antigens A and B.

5 People naturally have antibodies against A and B antigens if they do not have these antigens in their red blood cells. Hence, they do not have to be sensitized by previous contact with the antigens as is true of the other antigens that cause hemolytic disease of newborns.

6 Rh^D incompatibility results when the mother is Rh-negative and the fetus (and the father) are Rh-positive (Rh^D). Thus, the young lady who is about to be married is herself Rh-positive (Rh^D) and cannot produce antibodies against Rh^D. Thus, she will never be faced with the problem of bearing a child with Rh^D incompatibility.

7 This injected antibody can react with and destroy cells with the D antigen that might have entered the mother's body from the fetus. Hence, the fetal cells would be eliminated before they could cause the mother to produce any antibodies against this antigen.

8 You know that the allele for *A* was present because the A antigens are in the saliva. Also, the allele for *Se,* secretor, is present because the A antigen is in the saliva. Finally, the allele for the H substance is present because A antigen cannot be made without this substance.

9 Their chances for developing emphysema would be elevated somewhat, but not nearly as great as was the chance for emphysema in their father. He probably was homozygous for the gene for low activity of antitrypsin, while his wife was homozygous for the allele for high activity, so the children would be heterozygous.

10 It would be very difficult to extract proteins from the surface membranes of red blood cells to test by electrophoresis, yet the agglutination reaction to antibodies is simple. Conditions are reversed for the plasma proteins. The proteins are easy to separate by electrophoresis, but developing simple antigen-antibody reactions would be difficult.

11 She would be homozygous P^C/P^C, since each of these alleles contributes about 118 units of activity, higher than any of the other alleles.

Answers—Chapter Nine

1 An inborn error in metabolism is any disturbance in a normal biochemical pathway of an individual created by the complete absence or a deficiency of the particular enzyme that regulates a reaction in that pathway.

2 In *Galactosemia* the liver enzyme galactose phosphate uridyl transferase (GPT) is missing; in *alkaptonuria* homogentisic acid oxidase is lacking; in *phenylketonuria* phenylalanine hydroxylase is missing; and in *Tay-Sachs disease* hexosaminidase A is absent. In each of these cases (as well as others cited in the chapter), the disease is caused by the absence of a particular enzyme that is regulated by a single gene. Although many genes do regulate the assembling of amino acids into polypeptide chains that function as enzymes, other genes regulate the production of polypeptides that are not enzymes. Examples of non-enzymatic polypeptides are the α and β polypeptides of hemoglobin and the polypeptides of insulin.

3 Galactosemia is a genetic disease for which we have a treatment: dietary regulation permits normal development to occur. Tay-Sachs disease cannot be treated; it is fatal within the first three or four years of life.

4 The affected child born to a woman who has PKU and her "normal" husband is the victim of a harmful environment during the nine months of gestation. High levels of phenylalanine in the mother's body cause fetal damage and retardation. The mother has the genotype *pp*, her husband is in all likelihood *PP*, and their defective child is no doubt a heterozygote (*Pp*). If the pregnant woman had been placed on a low phenylalanine diet before she became pregnant (or at the very least, immediately upon becoming pregnant), the fetus may well have developed without the brain damage and other ill effects of environmentally induced PKU.

5 PKU is uniquely a genetic disease of whites; it is especially frequent in the Irish and their descendents. It is a rare disease among blacks, thus making screening for PKU in Washington, D.C. not very cost-effective.

6 Phenylalanine is an essential amino acid. We all need some phenylalanine in our diet; even children who have PKU must have some phenylalanine. The trick is to provide enough without providing too much. Careful monitoring of the PKU victim's diet is mandatory. Either an excess or a deficiency of phenylalanine can be damaging. On the other hand, lactose (milk sugar), the source of galactose, can be completely eliminated from the diet of a child with galactosemia, since lactose is not essential to normal development.

Answers—Chapter Ten

1 **(a)** replication (DNA reproducing DNA);
  (b) transcription (DNA regulating nucleotide sequence in messenger RNA);
  (c) translation (m-RNA regulating polypeptide assembly)

2 The DNA triplet CTT transcribes as the m-RNA codon GAA, which translates as glutamic acid; likewise, the DNA triplet CTC specifies GAG, which codes glutamic acid. Since GUA and GUG are m-RNA codons for valine (see Table 10–1), the following single nucleotide substitutions could produce the change from glutamic acid to valine in the sixth position in the β polypeptide:

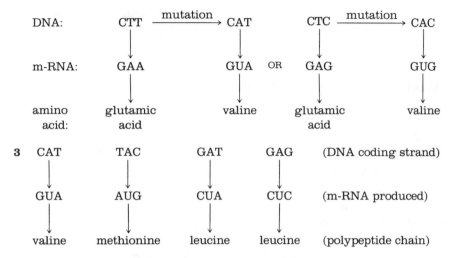

3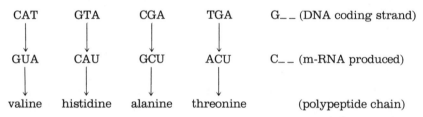

Now with the insertion of G after the CAT codon:

CAT	GTA	CGA	TGA	G_ _ (DNA coding strand)
↓	↓	↓	↓	
GUA	CAU	GCU	ACU	C_ _ (m-RNA produced)
↓	↓	↓	↓	
valine	histidine	alanine	threonine	(polypeptide chain)

Note that the amino acid sequence encoded by the mutated DNA will differ, beginning at the point of insertion of the nucleotide. Thus, the mutated polypeptide in our example differs from the original polypeptide after the first amino acid.

4 The frameshift mutation would be expected to be more serious in its consequences. The substitution of one base for another, as in the sickle hemoglobin mutation, results in the substitution of one amino acid for another in the polypeptide chain with all other amino acids being unaltered. A frameshift mutation alters all codons from the point of mutation on, resulting in numerous changes in the amino acids and a grossly altered polypeptide chain.

5 The substitution of A for G at one location in the second triplet would not alter the polypeptide chain coded by this gene. The original triplet, GGG, transcribes as the m-RNA codon, CCC, which is translated as the amino acid proline. The mutated triplet, GGA, is transcribed as CCU, which is also a m-RNA codon for proline. Thus, the mutation could be called a "silent" mutation because no change in the polypeptide product is produced and the mutation would not even be detected.

6 Since many of these chemicals are carcinogenic and teratogenic, they are suspected of being mutagenic. Most new mutations are detrimental or lethal.

7   The Huntington's disease gene is indeed lethal, but usually not until *after* the individual possessing it has married, reproduced, and transmitted the gene to his or her children.

Answers—Chapter Eleven

1   **(a)** $2n - 1$;   **(b)** $3n$;   **(c)** $4n$

2   **(a)** 47;   **(b)** 45;   **(c)** 69;   **(d)** 92

3   **(a)** 47, XXY;   **(b)** 46, XX, $5p^-$;   **(c)** 47, XX + 13;   **(d)** 47, XY, + 21;   **(e)** 45, XX, $t(14q^+, 21q^-)$

4   Banding techniques permit the detection of minute changes in chromosome structure that would not be visible under the microscope if the chromosomes were stained uniformly throughout their length.

5   The probability that a child from this mating will have Down syndrome is 1/2 or 50 percent. The probability that the offspring will be female who does not have Down syndrome is $1/2 \times 1/2 = 1/4$ or 25 percent.

6   **(a)** terminal deletion;   **(b)** intercalary deletion;   **(c)** isochromosome;   **(d)** translocation;   **(e)** acentric fragment (deletion of $ABCD \cdot EF$);   **(f)** inversion (of NOP to give PON)

7   Trisomy-21, translocation Down syndrome, and mosaic Down syndrome are the three types. Trisomy-21 accounts for about 95 percent of all cases. Mosaicism for trisomy-21 accounts for only 1 percent of the cases while about 4 percent of all cases involve a translocation. Mosaic Down is likely to be the least severe form since only a portion of the child's cells have the atypical chromosome number. The form of the syndrome most likely to recur in siblings of an affected child is translocation Down syndrome.

8   The risk of having a trisomic-21 child is relatively low when a woman is only twenty-five years old. The attending physician no doubt wished to determine whether the child has $2n = 47$, trisomy-21, or whether the cause of the condition is a translocation. If one of the parents is a translocation carrier, the couple is at high risk for having additional children with translocation Down syndrome.

Answers—Chapter Twelve

1   Contact inhibition means that when cells touch or contact one another in their growth they will cease growing. This is important because it stops the growth of the cells during repair when they contact one another. Otherwise, they would grow over each other and be malignant.

2   Cultured cells can be studied in detail in ways not possible for cells in the body. They can be experimented with in ways that could never be used on a living body.

3   Malignant cells lack contact inhibition and they do not become senescent. These are features needed for cultured cells. As we have discovered ways

to overcome contact inhibition and senescence in other cells, they have been used more widely.

4   Because we can grow human cells on selective media that will indicate only one mutant cell out of many nonmutant cells.

5   By removing the cells through amniocentesis and growing them on selective media that will reveal the presence of harmful mutations.

6   None, because hybrid cells of the mouse-human kind gradually lose the human chromosomes in repeated divisions.

7   When a normal cell is fused with a malignant cell, the result is nonmalignant as long as a particular chromosome from the normal cell is present.

8   We have to be able to recognize the particular part of a human chromosome that remains in hybrid cells that express a particular human characteristic and this was not possible before we identified the bands.

9   When cells from two slightly different mutants are grown close together they will be normal if the two genes are not alleles, but will be abnormal if the two genes are alleles.

10   We might take cells from the body of a person with a defect and have each of these cells take up a chromosome piece with a gene for the normal condition. Then these cells could be reinjected into the person with hope that they would function normally.

11   This is a judgment question. While putting their entire effort into the investigations, it would seem that discoveries would be accelerated, but the open exchange of information that characterizes university research would be lacking as each corporation tried to protect its discoveries as trade secrets.

12   Those hormones produced by human genes in bacteria would be exactly like those produced in the human body and would be more readily accepted and utilized by people who need the hormones. Also, hormones produced by bacteria would probably be cheaper to produce.

Answers—Chapter Thirteen

1   Those animals with a very short life cycle do not have time to learn the complex reactions necessary for existence and they must inherit them if they are to survive. With a longer life cycle there is time to learn and develop a better overall pattern of reactions.

2   Many different answers are possible, but any given should have logical justification. A typical answer might be the sucking response of babies. This is expressed in newborns who have had no chance for learning.

3   Same as Number 2. Typical answer might be driving a car. There is no instinct to press on the brake in an emergency; that must be wholly learned.

4 Assess the similarities and differences between monozygotic twins, dizygotic twins, nontwin siblings, and unrelated persons. This would give a good indication of the degree of hereditary influence.

5 The pattern is inherited in all people, but certain chemical changes in the body are necessary to bring it out. One gene can furnish the chemical.

6 Yes, because the ability to learn a language is innate and lacking one from adults, they would certainly develop one for communication.

7 Chromosome aberrations upset the gene balance and this is sufficient to alter the mental capacity, which depends on just the right amount of chemical and physical factors for proper function.

8 Persons reared as one sex or the other prefer to retain this sex even though they may have organs of the opposite sex.

9 No, they would not be influenced by parental dominance or other factors in our society that bring out attraction toward one's own sex.

10 A severe mental defect can be the result of homozygosity for a recessive gene and only one in four children would be expected to be homozygous if parents were heterozygous. Also, it may be due to a chromosome aberration that would not likely occur in more than one child. Mild retardation is more likely to result from polygenes and all children would share in these. Also, environmental forces causing mild retardation are likely to be shared by all.

11 Genes seem to determine the amount of stress that a person can stand without reaching the breaking point.

12 Heredity seems to determine potential, but environment is a powerful force necessary to bring out the potential. All can benefit from good education and training, regardless of heredity.

Answers—Chapter Fourteen

1 (1) (i); (2) (f); (3) (d); (4) (c); (5) (h); (6) (a); (7) (g)

2 Tuberculosis and measles are communicable diseases that are caused by a bacterium (tuberculosis) or a virus (measles). Nevertheless, the higher concordance for monozygotic relative to dizygotic twins suggests that a hereditary predisposition or susceptibility may also be a factor in the occurrence of these diseases. In the cases of spina bifida and clubfoot, the concordance values in monozygotic twins are decidedly higher than for dizygotic twins. This suggests that genes are important in the etiology of these conditions. The fact that concordance is considerably less than one for the monozygotic twins suggests, however, that these conditions are not exclusively genetic. In fact, most authorities regard them as polygenic traits involving a genetic threshold that must be exceeded before the conditions develop.

3 Incomplete penetrance complicates genetic analysis because an individual's phenotype may not be an accurate reflection of his or her genotype. A dominant trait, for example, may appear to "skip" a generation because of incomplete penetrance.

4   Heredity; environment

5   penetrance

6   Environmental factors that interact with the genotype or perhaps mod-
    ifying genes that prevent the expression of the major gene for the trait
    may both play a role in causing incomplete penetrance.

7   Many different carcinogens may function as mutagens; i.e., they may
    produce cancers by damaging DNA, so that the normal processes of reg-
    ulation of cell division are damaged or destroyed.

Answers—Chapter Fifteen _____

1   Apparently we have an example of the founder principle. A sampling
    error may have occurred such that the original settlers in Indiana were
    not a representative sample of the population from which they came in
    that the settlers included a disproportionately high frequency of HD vic-
    tims. This disproportionate frequency has persisted to the present.

2   Many man-made environmental pollutants may have a mutagenic effect,
    and medical X rays, atomic fallout, and other forms of radiation all tend to
    increase the number of new mutations in the human gene pool. Many
    advances in medicine make possible the survival of individuals possessing
    a variety of genetic defects that would have been lethal in earlier times.
    These individuals can thus reproduce and further increase the number of
    deleterious genes in the gene pool.

3   Many possible examples: Eyeglasses permit the nearsighted to survive,
    reproduce, and transmit their genes for poor vision. Insulin for the dia-
    betic, antihemophilic factor for the hemophiliac, low phenylalanine diet
    for the PKU victim, and numerous other developments also permit sur-
    vival to reproductive age and consequently produce a dysgenic effect.

4   The population must mate randomly with respect to the trait in question.
    Furthermore, neither the A nor the a allele can have a selective or muta-
    tional advantage. (That is, $A/A = A/a = a/a$ in terms of surviving and
    reproducing, and mutation must be absent or $A \longrightarrow a$ must equal
    $a \longrightarrow A$.) Likewise, migration and genetic drift must not alter the allele
    frequencies in the population.

5   If 16 percent are Rh-negative, then $q^2 = 0.16$, $q = \sqrt{0.16} = 0.4$, and $p =$
    0.6. It follows, then, that the frequency of $A/A = p^2 = (0.6)^2 = 0.36 = 36$
    percent and that the frequency of $A/a = 2pq = 2(0.6)(0.4) = 0.48 = 48$
    percent.

6   Since 9 percent of the population is a/a, it follows that $q^2 = 0.09$ and
    $q = \sqrt{0.09} = 0.3$. The value of $p = 0.7$ and $p^2 = (0.7)^2 = 0.49 = 49$ percent
    A/A and $2pq = 2(0.7)(0.3) = 0.42 = 42$ percent A/a. Since the frequency of
    A/A in the population under consideration is 33 percent and the frequency
    of A/a is 58 percent (quite different frequencies from what was calculated
    on the basis of $q = 0.3$ and the Hardy-Weinberg equilibrium), we must
    conclude that the population is *not* in equilibrium.

7 $q^2 = 1/40,000$ and $q = \sqrt{1/40,000} = 1/200$. The value of $p = 199/200$. Therefore, $2pq = 2(199/200)(1/200) = 398/40,000$. Thus in 40,000 individuals, one would expect to have 398 T/t. This is an approximate frequency of 1 percent: 398/40,000 is approximately equal to 400/40,000 = 4/400 = 1/100 = 1 percent.

8 At the present time $q^2 = 1/10,000$ for PKU and $q = 1/100$ (i.e., $q_0 = 1/100$). After ten generations of complete selection against PKU by sterilization,

$$q_{10} = \frac{q_0}{1 + nq_0} = \frac{1/100}{1 + 10(1/100)} = \frac{1/100}{11/10} = \frac{1}{100} \times \frac{10}{11} = \frac{1}{110}$$

It then follows that the frequency of PKU victims after ten generations of selection will be $(q_{10})^2 = (1/110)^2 = 1/12,100$. Thus, where now 1 in 10,000 white babies is born with PKU, in ten generations 1 in 12,100 would have PKU—hardly a marked reduction. Selection has relatively little effect here because the PKU gene is rare and recessive. Selection against it is prevented because it is so often found in heterozygous individuals.

Answers—Chapter Sixteen

1 (1) (h); (2) (f); (3) (c)

2 family physician, medical schools within state, state medical societies, professional directories of genetic counseling services

3 Many people who have had a defective child have feelings of guilt, anger, disbelief, and depression. Furthermore, the couple may reject the defective child or the husband may blame the wife (or vice versa) for the child's defect. In other situations serious moral and ethical problems may be raised by the options opened to the counseled couple. In all of these cases a counselor who is specially prepared to face moral-ethical issues and the trauma of having produced a seriously defective child can be of great help to the counselees.

4 Amniocentesis with prenatal detection of defects, artificial insemination using donor semen (AID), and modern methods of contraception are three obvious examples.

5 Answers will vary, but most of us would want a genetic counselor who is knowledgeable of the current developments in human and medical genetics, who has access to excellent laboratory facilities and consultants, and who has the ability to empathize with the counselee.

index

Genetics and biology textbooks published in the past have frequently included, as examples of single gene traits, many conditions of questionable authenticity. To prevent promulgating such misinformation, the authors have attempted to document each genetic condition, disease, or defect discussed in this book by identifying its catalog number in V. A. McKusick's *Mendelian Inheritance in Man*, Fifth Ed. (1978). Thus, the five-digit number in parentheses immediately following an entry in the index is the McKusick catalog number.

Cancer (continued)
  therapy agents, 38
  in twins, 240
Carcinogenic agent, 242, 247–49
Carcinomas, 242
Carriers, detection of gene, 56–57
Catatonic withdrawal, 236
Cavalli-Sforza, L., 223, 230, 274
C-bands, 12
Cellular immunity, 109
Center for Education in Human and Medical
    Genetics, 277, 278
Centromere, 11
Childs, Barton, 279, 285
Chondrodystrophic dwarfism (11865), 51,
    167, 168, 169
Chorion, 43
Christmas disease. See Hemophilia B.
Chromatids, 11
Chromosome aberrations, 175–95, 245–46
Chromosome polymorphism, 44
Chromosomes, 7–14
  bands on, 12–14
  sex, 71–78
Chronic granulocytic leukemia. See Chronic
    myeloid leukemia.
Chronic myeloid leukemia (CML), 190–91
Cigarette smoke, carcinogenic effects of, 245,
    248
Cleopatra, 272
Clonal selection theory of antibody
    production, 112
Clubfoot (11980), 240
Codominant inheritance, 56
Codons, 160, 163–64
Colchicine, 10, 186
Colorblindness, 91, 92–95
  deutan (deuteran) type (30380), 93
  protan type (30390), 93
  tritan type (30400), 93–94
Complement, 137
Complementation test, 208–10
Concordance, 238–41
Confidentiality of medical records, 281
Conjoined twins. See Siamese twins.
Consanguineous matings, 272–73
Contact inhibition, 199
Contraception, 283
Cooley's anemia (27350), 258, 284
Cortisol, 152
Cortisone, 38
Cowie, V., 227, 230
Cretinism, genetic goiterous (27440–27490),
    149–50
Crick, F. H. C., 2, 16
Cri du Chat syndrome, 189–90
Crossing over, 187
Cry of the cat syndrome, 189–90
Cultural selection, 255–56
Curie, Marie, 248
Cyclamates, as cancer cause, 247
Cystic fibrosis (21970), 40, 57, 260, 261, 282,
    284
Cytomegalic inclusion disease, 39
Cytosine, 3

Daughter cells, 4
Deafness, 67–68, 240
Defective dentine, 98
Deletions, 187–88
Delta chains of hemoglobin, 162
Deoxyribonucleic acid. See DNA.
Dermatoglyphics, 44
DES, 38
Deutan type of colorblindness. See
    Colorblindness.
Deuteran type of colorblindness. See
    Colorblindness.

Deuteranomaly, 93
Deuteranopia, 93
Diabetes, 115, 118, 240, 262
Diagnosis of genetic defects, 280
Dichorionic twins, 43, 44
Diego blood antigen (11050), 135
Diethylstilbestrol. See DES.
Directories of genetic counseling services,
    279, 285, 286
Dizygotic twins, 44
DNA, 2–4
Dombrock blood antigen (11060), 135
Dominant, 50
Donor semen, 282
Down syndrome, 41, 46, 176–82, 193
  cancer in, 246
  in twins, 240
Drift, genetic, 256–57
Drosophila, 170
Duchenne muscular dystrophy. See
    Pseudohypertrophic muscular dystrophy.
Duffy blood antigen (11070), 135
Dunker, blood types, 257
Dwarfism, pituitary (26240), 153

Edwards syndrome, 182–83
Egg, 17, 25
Ehlers-Danlos syndrome (13000), 61–62
Ehrlich, P., 223, 230
Ehrman, L., 219, 230
Electrophoresis, 137, 138, 163, 200
Embryo, development of, 34–36
Emotional overtones of genetic defects, 281–
    82
Endonuclease, 243
Endorphins, 217
Enkephalin, 217
Environment, related to heredity, 231–51
Enzyme deficiency diseases, 143–52
  control of, 152–53
Enzymes, as gene agents, 143–57
Epidermolysis bullosa (22660), 169
Epilepsy, 240
Epiloia (19110), 169
Epistasis, 66–68
  of blood groups, 128–29
Epsilon chains of hemoglobin, 162
Equilibrium, Hardy-Weinberg, 264–73
Erythroblastosis fetalis, 130
Erythrocyte acid phosphatase (17150), 136
Erythrocytes. See Red blood cells.
Estrogens, 38
Euchromatin, 12
Eugenics movement, 228, 283
Expressivity, variable, 234–35
Eye color (22724), 65
  selection for, 270–72

Fallopian tube, 29, 30
Family physician, 278–79
Fanconi's anemia (22765), 244
Favism (13470; see also 30590), 98–99
Feldman, S., 223, 230
Fertilization, 29–32
Fetal alcohol syndrome, 38, 39
Fetus, 36
Fibrinogen, 139
Fibroblasts, 199
Fiji, populations in, 259
First polar body, 25
Fisher, R. A., 133
Flemming, Walther, 8, 9
Flipper arms (10790?; 17910?), 234
Fluorescent Y-chromosomes. See
    Y-chromosome.
Fölling, A., 146, 156

Kallman, F. J., 221
Karyotyping, 11
Kell-Cellano blood antigen (11090), 135
Kidd system (11100), 135
Kidney transplants, 7, 116
Killer cells, 109
Klinefelter syndrome, 73–74, 76, 78

Lacks, Henrietta, 199
Landsteiner, Karl, 124, 130, 135, 141
Lead, 38
Lesch-Nyhan disease (30800), 99, 201, 202, 217, 226
Lethal genes, 57–59
  expression in consanguineous matings, 273
Lethal mutations, 165
Leucocyte drumsticks, 80
Leukemia, 242
  chronic myeloid, 190–91, 246
  in Down syndrome, 246
Levan, A., 9
Lewis blood antigen (11110), 135
Li, C. C., 259, 275
Lincoln, Abraham, 61
Linkage, 62
Lithium, 38
Locus, 49
LSD, 38, 192
  mutagenic effects of, 170, 192, 193
  teratogenic effects of, 38
Lupus. See Systemic lupus erythematosus.
Lutheran blood antigen (11120), 135
Lymphocytes, 107, 199
Lyon, Mary, 80
Lyon hypothesis, 80, 100
Lysins, 109
Lysis, cell 109
Lysosomes, 108

Macrophages, 108
Malaria, resistance to, 99, 258
Malignant cells, 199
Malignant neoplasms, 242
Mammographs, 248
Maoris, 256
March of Dimes Birth Defects Foundation, 41, 277, 278
Marfan syndrome (15470), 59–61, 169
Marihuana, 170
Marks, J., 278, 286
McKusick, V. A., 104, 204, 206–7, 214
Measles, 240
Medical school, 279
Meiosis, 22
  meiosis I, 23
  meiosis II, 23
  sex chromosome, 17, 72, 92
Meiotic drive, 181
Melanin, 50, 150
Melanoma, 242
Memory cells, 109
Mendel, Gregor, 49, 69
Mendel's laws of heredity, 49–63
Mental retardation, 144, 147, 225–29, 240
Mercury, 38
Messenger-RNA (mRNA), 5, 6, 15
Metacentric centromere, 11
Metaphase, 19
Micropthalmia (15685), 169
Microwaves, 38
Migration, gene flow through, 258–59
Mikamo, K., 86
Mills, Ivor, 103
Mitochondria, 24
Mitosis, 17–19
MN antigens (11130), 134–35, 265–66
Mongolism. See Down syndrome.

Mongols, invasion of Europe, 259
Monochorionic twins, 43
Monosomic, 176
Monozygotic twins, 42
Morton, N. E., 230, 261
Morula, 33
Mosaicism, 185–86
Mosaics, sex-chromosome, 78
Muller, H. J., 31, 168, 173
Multiple alleles, of blood groups, 126
Multiple births, 42–47
  sex ratio in, 86
Multiple polyposis (17530 and others), 169, 243
Multiple sclerosis, 114, 119
Mumps, 39
Muscular dystrophy (31020), 96, 169, 284
Mutagenic agents, 167–71
  carcinogenic effects of, 249
  in environment, 269
Mutations, 5, 159
  detection of, 170–71
  differential, 257–58
  frequency of, 167
  harmful, 268–69
  kinds of, 164–67
  rate of appearance, 169
  X-linked lethal, 171
Myasthenia gravis (10910; 25420), 114
Myocardium, 114
Myoma, 242

Nail-patella syndrome (16120), 169
National Clearinghouse for Human Genetic Diseases, 277, 278
Natural selection, 253–54
Navaho Indians, albinism in, 150
Neoplasm, 242
Nephritis (16190), 114
Nondisjunction, of chromosomes, 73, 175, 177–78
Nonprescriptive genetic counseling, 280
Nonsecretors. See Secretors.
Nonself, 112–13
No-X egg
  fertilized by an X-sperm, 74–76
  fertilized by a Y-sperm, 76
Nuclear
  generating plants, 169
  weapons, 169
Nucleases. See Restriction enzymes.
Numerical chromosome aberrations, 175–87

Oncogene theory of cancer induction, 247
Oncogenic agent. See Carcinogenic agent.
Oocyte, 25
Oogenesis, 24–25
Ootid, 25
Osteogenesis imperfecta (16620 and others), 234–35
Ovulation, 29–30

Painter, T. S., 8, 9
Paralytic polio, 115
Paranoia, 236
Parsons, P., 219, 230
Parthenogenesis, 33, 47
Patau syndrome, 183, 184
Pattern baldness (10920), 101, 103
Pedigree, 280, 281
Penetrance, reduced, 235
Penicillin sensitivity, 106
Pernicious anemia (24020), 115
Pesticides, 39
Phenocopy, 232–33
Phenothiocarbimide. See PTC.
Phenotype, 50

Phenylalanine, 146
Phenylketonuria. See PKU.
Philadelphia chromosome, 190
Phocomelia (26900; 27400), 37, 232, 233
Pima Indians, diabetes in, 262
Pituitary dwarfism. See Dwarfism.
PKU (26160), 146–49, 206, 208, 226
  gene frequency, 266
  in monkeys, 233
  prevention of, 154
  variable expressivity of, 234
Placenta, 34
Plasma cells, 108
Plasma proteins, 137–40
Plasma thromboplastic component (PTC). See
    also Hemophilia B., 95
Pleiotropy, 59
Polar body, 25
Polygenes, in cancer, 244
Polio, 115
Polydactyly (17420 and others), 235, 236
Polyethylene glycol, 202
Polygenic inheritance, 63–66
Polymorphism
  balanced, 262–63
  of blood groups, 263–64
  of gene pools, 261–63
  transient, 262
Polypeptide chain, 5
Polyploid, 186–87
Population, 253
Population genetics, 253–75
Precipitins, 109
Pregnancy, tubular, 32–33
Prenatal detection of mutant genes, 202
Prescriptive genetic counseling, 280
Primary
  oocyte, 25
  spermatocyte, 23
Probabilities, genetic, 51–53
Prophase, 18
Prospective genetic counseling, 279
Prostaglandins, 30
Protanomaly (30390), 93
Protanopia (30390), 93
Protan type of colorblindness (30390), 93
Provirus, 246–47
Pseudohemophilia (13450; 14232), 95–96
Pseudohermaphrodite (26427), 86
Pseudohypertrophic muscular dystrophy
    (31020), 96–97, 169
PTC, inheritance of the ability to taste
    (17120), 264
Ptolemaic dynasty of Egypt, 273
Puck, Theodore, 10, 191, 214
Pyloric stenosis (26600), 261

Q-bands, 12, 13
Quinacrine mustard, 12

Race, R. R., 133, 141
Radiation, ionizing, 96, 168–69
  mutagenic effects of, 269
Radioactive isotopes, 168
Ratios, genetic, 51–53
Rauscher mouse leukemia virus, 246
R-banding technique, 12
Recessive, 50
Recombinant DNA, 155, 210–12
Red blood cell antigens, 54, 56, 123–36
Reduced penetrance, 235
Reflexes, 216
Replication, 4–5
Reproductive cells, formation of, 22–25
Reproductive selection, 255
Restriction enzymes, 212
Retinoblastoma (18020), 169, 243

Retrospective genetic counseling, 280
Reverse transcriptase, 246
Reye syndrome (see 31125), 114
Rh antigens (11170), 129–34
Rheumatic heart disease, 113–14
Rheumatoid arthritis, 114
Rh factor. See Rh antigens.
Rh incompatibility, 130–33
Rh-negative, 130
Rhogam, 132
Rh-positive, 130
Ribonucleic acid. See RNA.
Ribosomes, 5, 6
Rickets
  vitamin-D-resistant (30780), 240
  in twins, 240
Ring chromosomes, 188
Risk factor, 280, 281
RNA, 5–6
Rous sacroma virus, 246
Rubella, 37, 39, 192
Ruddle, F., 204, 214

Saccharin, carcinogenic effects of, 248
Sampling error, 256
Sandberg, A. A., 76
Sarconomas, 242
Satellites, 12
Schizophrenia (18150), 236–37, 240
Scoliosis, in Marfan syndrome, 60
Secondary oocyte, 25
Secondary spermatocyte, 23
Second polar body, 25
Secretors of the A, B, H antigens (18210), 129
Segregation, 51
Selection, effects of, 270–72
Selective medium, 200, 204
Selenium, 38
Self, 112–13
Semen, 30
Sendai virus, 202
Serotonin, 237
Serum, blood, 124
Sex
  abnormalities of, 73–78
  determination of, 71–89
  differentiation, stages of, 83–84
  dimorphism, 263
  hormones, 82–86
  influence on heredity, 91–104
  ratio, 86–88, 96–97, 171
  reversal, 83
Sex-chromatin body. See Barr body.
Sex chromosome mosaics, 78
Sex chromosomes, 71–78
  identification in newborns, 79, 81
  identification in women athletes, 82
  synapsis of, 23, 92
Sex-influenced traits, 101–2
Sex-limited genes, 102–3
Sex-linked genes. See X-linked genes.
Sexual behavior, 220–21
Sexual dimorphism, 263
Siamese twins, 46–47
Siblings, 238
Sibs. See Siblings.
Sickle-cell anemia (14190), 54–55, 258, 260,
    266, 282, 284
  in blacks, 260, 261
  selective advantage of, 266
  test for carriers, 270
Sickle-cell hemoglobin (14190), 163
Sickle-cell trait, 54
Simian virus, 40, 204–5
Skin graft test, 45
Skin pigmentation, 65–66, 232, 254
Skip-generation method of inheritance, 91